SACRED GROUND

How Climate Change is Devastating
America's National Parks

FREE GREENMAN

BALBOA
PRESS
A DIVISION OF HAY HOUSE

Copyright © 2019 Free Greenman.

All rights reserved. No part of this book may be used or reproduced by any means, graphic, electronic, or mechanical, including photocopying, recording, taping or by any information storage retrieval system without the written permission of the author except in the case of brief quotations embodied in critical articles and reviews.

Balboa Press books may be ordered through booksellers or by contacting:

Balboa Press
A Division of Hay House
1663 Liberty Drive
Bloomington, IN 47403
www.balboapress.com
1 (877) 407-4847

Because of the dynamic nature of the Internet, any web addresses or links contained in this book may have changed since publication and may no longer be valid. The views expressed in this work are solely those of the author and do not necessarily reflect the views of the publisher, and the publisher hereby disclaims any responsibility for them.

The author of this book does not dispense medical advice or prescribe the use of any technique as a form of treatment for physical, emotional, or medical problems without the advice of a physician, either directly or indirectly. The intent of the author is only to offer information of a general nature to help you in your quest for emotional and spiritual well-being. In the event you use any of the information in this book for yourself, which is your constitutional right, the author and the publisher assume no responsibility for your actions.

Any people depicted in stock imagery provided by Getty Images are models, and such images are being used for illustrative purposes only.
Certain stock imagery © Getty Images.

Print information available on the last page.

ISBN: 978-1-9822-1944-4 (sc)
ISBN: 978-1-9822-1956-7 (e)

Balboa Press rev. date: 01/10/2019

DEDICATED TO SALLY

Thanks for providing the necessary space and time needed to birth this book. My message to the world was made possible by your vast generosity, for even at 84 years old, you never forgot how to give with an absolutely open heart.

OPENING STATEMENT

Climate change is a controversial topic for many people. I felt compelled to write about it anyways due to my grave concern for what is happening in our National Parks. Excess carbon dioxide swirling in our atmosphere is heating our planet unnaturally fast and the impacts are being felt like a wave in one National Park after the other across our great country. Species are being squeezed by biological pressure cookers that could unravel ecosystems around the globe.

It is time to examine our role in this global catastrophe for the wrongdoings we commit against nature are also becoming crimes against humanity, returning to us tenfold, circling around like a boomerang. If we don't halt the direction we are headed with our carbon dioxide emissions, human suffering will be as immense as the rising seas, as disastrous as rivers of rain flooding our cities and as scorching as summer droughts withering crops desperately needed by hungry people. As it turns out, our changing climate is as much a social issue as an environmental problem impacting our parks.

In writing about these topics, I operated solely as an independent person, not as a representative, nor spokesperson, of any agency. Though I have kicked around my share of parks as a visitor, the thoughts contained herein are my own, not necessarily those of the National Park Service or any other agency. This is my personal story of concern. I felt duty-bound to share it in the hope of encouraging immediate change, for the betterment of all American families who have ever frolicked, played and spent themselves in a wild and beautiful park, creating memories that last a life time and for the sacred landscapes themselves which contain so much of what is raw and inspiring in our great country. These wild open spaces lack a voice to utter their own disquiet regarding the unfettered changes roiling across their ridgetops at break-neck speed.

I wanted to be that voice, on behalf of these landscapes, so a tiny fraction of their heart-felt stories could be told. Understanding the plight of these magnificent ecosystems also foreshadows the struggles we will all soon face as our physical world unravels due to our rapidly changing climate. It is time for Americans to comprehend what's at stake. Denying the changes being wrought by our faltering climate has set ill-fated wheels in motion that could prove disastrous not only to our most sacred places on earth, but for all of humanity.

Buckle in, the problem is far more pervasive and devastating than you realize.

A CALL TO ACTION

Invisible, natural forces have altered our planet's climate over the millennia. Just consider that the Petrified Forest in parched Arizona was once dripping with tropical plants and roving dinosaurs that stripped leaves from trees as they stomped through murky swamps.

Hidden beneath the surface of our planet are physical, chemical and ecological forces that drive our climate. Melting ice caps, both at the poles and in the tropics, play a role. So does the amount of fresh water pouring off of melting glaciers for salt water is heavier than fresh. When the ratio is in balance, salt water sinks, allowing it to plummet through the ocean's depths where it glides along the seafloor, driving major currents of water to flow between the poles.

Ice ages have, in part, been driven by low levels of carbon dioxide, running around 180 parts per million whereas warmer spells were swarming with more CO_2 in the atmosphere, ranging in the realm of 240 to 280 ppm instead. Burps of methane emitted from huge craters on the oceans floor could partially be responsible for some of the natural climate shifts. As these pockmarked craters historically released methane into the sea, the gas grabbed oxygen molecules on its way to the surface, forming carbon dioxide. As this CO2 bubbled into the atmosphere, it created an insulating layer in the air, bouncing heat back onto the planet thus warming the earth.

Solar flares, earth wobbles, the elliptical path of our planet around the sun and even the amount of verdant plant life are a few more of the unseen forces that have controlled and driven our climate since time began.

Changes in the climate are natural, though they have not always proceeded at a steady, incremental pace, much to our consternation. Sudden shifts between calm states have been witnessed in ice cores dug from deep within the snowfields of the polar regions, swinging wildly before settling into less turbulent rhythms.

What's unique about our changing climate today is that the speed of change has accelerated to a startling degree. During the retreating ice age of 20,000 to 10,000 years ago, temperatures soared about 9 degrees Fahrenheit in a ten thousand year stretch, exhibiting a seemingly rapid, yet natural shift in climate. By comparison, the climatic changes unfurling on earth now are happening about 30 times faster, leaving wildlife, plants, butterflies, krill and migrating birds scrambling to adjust. It is this very speed and ferocity of change that has scientists gravely concerned, for it is anything but natural.

One of the few changes between then and now is that humans have colonized the earth. By altering the surface of the globe to suit our needs, humans have distorted the physical machinations that once ran the planet's climate. Large-scale farming, deforestation, mining and burning fuel needed to spur our industrial revolution are the primary factors that have changed since the last ice age retreated to the glaciers of the far north.

All of these activities release vast amounts of carbon dioxide. In fact, so much CO_2 is getting pumped into our air that our oceans are sucking up tons of it just to try and keep the peace. By increasing carbon dioxide in our air, we have not only warped time scales but we have also changed the rules of the game by altering our own atmosphere. As we continue down this unbridled path, from sea to shining sea, Americans are paying the price for not recognizing the role we play in creating long-term climate trends that veer far from the natural cycles once set in motion by a higher hand and the physics of the spheres.

Our shortsightedness and greed have purchased us faucet water that we can literally light on fire, air too filthy to breathe, flooded streets in Miami, and human induced earthquakes caused by forcing millions of gallons of chemical-laden wastewater down fracking wells so we can have a "green transition fuel" that is not "harmful" to the environment. Meanwhile, ecosystems all around the globe are collapsing as our climate changes at an unnatural pace.

The impact to America's National Parks is also astounding. No matter which park you visit, a distressing story emerges regarding how climate change has impacted resources at an unprecedented speed and scale.

Resource damage of this temerity crosses my line in the sand. America's National Parks mean everything to me. They have been my solitude and inner peace, my sweaty joy and perseverance in the face of vast and daunting odds, as well as my reason for being. They have also become my church, my place of worship, for it is here that I see God's hand so plainly. I did not so much choose the parks as they chose me. Guess I have a penchant for visiting the finest outdoor churches in the world. Looking back over the great luck of my life, I have been blessed to scale outrageous mountains and float down winding, graceful rivers that slice through stone and time. Having received so much joy from these wild and gritty places, it is now time to give back.

From Zion to Yellowstone and the sawgrass prairies of Everglades, National Parks are America's Best Idea because they capture the essence of all that is beautiful in our country. Though they too are faltering with the onslaught of changes and outside influences battering at their doors, they still provide a glimpse into ecosystems striving to maintain balance. The average person can enter a park and be enthralled with the scene displayed before their eyes. Knocked to their knees with gratitude for all that is and thankful that we had the wisdom early on to set such pristine places aside.

Bighorn sheep can be witnessed butting heads during the fall rut, the clacking sound of their horns reverberating through ice-cold air, while steam rises fog-like from their nostrils. The eerie howl of wolves echoes across great lakes where the boreal forest dips down along rocky shorelines. Forming a backdrop to this wolf music are the northern lights, an ethereal curtain of green shimmering across the night sky.

Petroglyphs provide glimpses into other cultures, reminding us that there are other ways of being, living a simpler more connected life. A life where we paid attention to the stars, the cycles of the moon and the growing of crops.

Wild and reckless rivers gush through narrow canyons reminding us that not everything in nature has been tamed, nor is it meant to be. For in the taming of such a wild thing, you crush its spirit, its very essence, its reason for being.

Our National Parks mean everything to me. They are a place of solace, of healing, of connecting to God through the beauty of nature. They are also in serious trouble.

Though I know many people worship differently than I, most people who enter church have a feeling of sacredness and awe coupled with a desire to be whole. Church should be a place of safety and sanctity, of pureness, a place that holds the feeling of "Everything will be alright no matter what goes down."

This is no longer the feeling I get when I enter my places of worship. Instead, I see places of divinity in peril, where catastrophic changes are happening at rapid-fire speed on a grand scale, where millions of years of evolution are being trifled with casually rather like an accidental splitting of atoms. I find myself wondering a lot lately when we will hear the irreversible boom, the moment we recognize our own death, the flash we see when we realize what we've done can't be taken back.

In short, America's National Parks are facing cataclysmic change. Much of that change, which is occurring on an ecosystem level, is perpetrated by our rapidly changing climate.

I'm not here to convince you that humanity is causing this change, that is for you to determine. My purpose is to goad you to investigate, to read, to weigh the evidence and to consider the ramifications, moral and otherwise. Educate yourself on the matter and decide what is right, then make a stand for yourself for this is your America too. How we choose to make it great is a uniquely individual pursuit, but if like-minded people band together we have a more powerful and collective voice in the matter. First, however, we must understand the forces that drive nature.

Visiting a park without learning about these forces leaves one empty of understanding and riddled with questions such as: How is balance kept amongst elk herds when a key predator such as the wolf has been systematically removed from the picture? It takes thousands of years of evolution for an apple snail to learn where to lay its eggs on a blade of sawgrass, so what happens when water levels unexpectedly rise past historic levels? What character traits did the people on Flight 93 have that enabled them to stop a hijacked plane from dive-bombing our nation's capital?

How can you connect to a place if you don't understand its underpinnings or how it works? What are the emotions or heart center of a place? What are its tastes and smells and gritty feel? How does it tick? When you visit an historic area, do the voices of the past reach out and grab you viscerally? If while on a cruise, you see a huge chunk of ice calve off a glacier, do you wonder why?

Park Rangers serve as a voice for these landscapes so their truest, most heart-felt stories can be told. In hearing these stories a portal is opened and visitors begin to get the true meaning and significance of a place. Deep and abiding connections are made once people choose to walk through that portal and that creates a sense of caring and stewardship.

I also like to tell stories about nature and you cannot separate park landscapes from our ever-changing climate. That's why climate change is one subject we can't afford to be ignorant of, it means too much to too many special places. It is impacting every acre of land entrusted to the National Park Service's care. If you are concerned about your National Parks, or even the beauty and sanctity of your own backyards, please take the time to make an educated decision for yourself… but read quickly, time is of the utmost importance. This problem is not new, it has been going on my entire life. It is well entrenched, ongoing, overarching and pervasive. Very likely, it is the biggest threat our National Parks will ever face.

Since 1963, scientists started warning our country's leader about the perils of climate change. It all began with a man named Charles Keeling who was measuring carbon dioxide levels from the Mauna Loa Observatory in Hawaii. Year by year he took measurements which showed an obvious upwards spiral from 1958 on. By 1963, Keeling had seen enough to raise a warning flag in the form of a letter he sent to President Johnson, alerting him to the steady, unnatural rise in the CO_2 levels he was witnessing.

Mauna Loa Observatory turns out to be a good place to collect such data for it is perched in the middle of the Pacific Ocean well away from factories and forested lands that could artificially boost measurements. Still, the curve shows powerful annual cycles that can't be overlooked. In the spring and summer, zigzags on the curve are visible, showing drops in the CO_2 level as plants in the northern hemisphere come alive and begin using carbon dioxide to grow, capturing and sequestering the carbon within their tissues. In the winter, when plants are more dormant, the gas levels rise once again, showing the cyclic breathing of our planet.

More recently, research has shown that pre-industrial levels of CO_2, between the years 1,000 and 1,750 AD, ranged between 275 and 285 parts per million. By 2005, we had zoomed to 380 ppm, with an upward trend of about 20 ppm predicted per decade, however, by Dec. 28, 2017, we had already hit the 408-ppm mark. It's been several million years since we have catapulted past this threshold and that concerns scientists greatly.

As climate change scientist James Hansen once said, "If humanity wishes to preserve a planet similar to that on which civilization developed and to which life on Earth is adapted, CO_2 will need to be reduced to at most 350 ppm…" though lower than that is ideal.

We have already sailed well beyond the 350 mark. Turning back the tide will take a concerted effort, requiring everyone's awareness coupled with a call to action to create positive change. Continuing to overshoot recommended CO_2 levels could unleash irreversible effects around the globe. At this juncture, from a risk management perspective, should we busy ourselves pointing fingers at the cause or do everything within our power to simply lower the CO_2 levels now? What, after all, is at stake?

We MUST decide, as a species, what we want to leave behind. As conscious, moral beings we are being forced into box canyon so tight we cannot see our boots or even a clear path forward.

In many parts of the west and Alaska, fires are getting bigger and more intense. The term megafire has been coined to capture the catastrophic nature of these events. Elsewhere, glaciers are melting at an unprecedented pace. Rising sea levels are impacting many coastal towns and villages and species of all sorts are trying to make the shift from where they once called home, to somewhere new they can tolerate.

Virtually no aspect of the parks will remain untouched. From the perspective of one who worships within these sacred cathedral walls and rushing rivers, I am afraid for all species, including us.

IN THE REALM OF GIANTS

On December 31, 1999, I crawled into a fire-scar in the base of a Sequoia tree in Giant Forest and prepared to greet the next century in ranger-like fashion. After unrolling my sleeping bag, I lit a candle and watched swirls of mist form overhead. I chanted and sang songs, shivered in the cold till my teeth rattled and admired the glow of light as it flickered across my tree den. Sequoia trees are magical and I couldn't think of a finer place to celebrate the dawning of a new age than the Room Tree in Sequoia National Park.

Despite the solitude and peace I experienced inside my tree, many tree species including Giant Sequoias and Coastal Redwoods are in the process of adapting to rapid-paced change, long periods of drought being one of them. Following my 4th year of extended drought, I left California after the Rim Fire swept through central California, becoming the third largest fire in the state at that time and one that was certainly spurred on by brittle dry conditions.

The Rim Fire, which consumed more than 257,000 acres of timberland, did not actually go out for over one full year due to residual heat pockets that burned away in stumps and roots over the winter, well into the heart of the following fire season. While this was taking place, water levels dropped dangerously low in reservoirs that fed water to my sink. From my perspective, it was time to move out of the state before there was no water left to drink. Meanwhile, deep-rooted blue oaks, stalwart ponderosa pines and even hardy manzanita bushes, with their red twisty branches, began dying off in vast numbers in the foothills around me.

Having lived for thousands of years, our nation's Big Trees have faced hardships before, some of them dating back to Medieval times. And though many climate scientists would agree that our changing weather patterns are potentially disastrous, it does appear that excess atmospheric carbon is spurring the growth rates of these trees in this century.

Trees draw in carbon dioxide through tiny holes in their leaves called stomata. Once in their system, trees break down the carbon into its useful parts, using some of them to put on girth and others to create the sugary sap that feeds them. With a huge surplus of carbon dioxide in the air, Big Trees are increasing their ring wood production at an unprecedented rate during this century. By capturing, or sequestering this excess carbon, our most majestic trees are helping to lower unhealthy rates of the greenhouse gas in the atmosphere.

Sequestering works for a while, until a drought strikes and the trees and underlying soil burn, leaving behind a charcoal forest of burnt sticks. Then instead of acting as a carbon sync, massive volumes of the gas are released back into the atmosphere via huge dominating plumes of smoke that swirl above burning forests.

In the meantime the trees, like our oceans, are providing us some measure of protection by storing away the excess carbon emitted into our atmosphere. In places where Redwoods grow, some of the highest amounts of stored, above-ground carbon on earth can be found, socking away more than 5,511,557 pounds of carbon per each 2.5 acres of tree cover. The excess carbon is actually enabling these trees to become more water efficient by tightly regulating the opening of their stomata which in turn stops them from losing moisture through their breathing-transpiration process.

That is not to say that too much carbon is a good thing for it has varying effects on the environment of these trees. Hotter, drier conditions create water stress and this is easier for adult trees to handle than young seedlings. If excess carbon continues to heat things up, which seems to be a certain trend at present, the youngest giants may pay the price for they have not yet built the extensive root systems needed to capture soil moisture during the driest of times. Mature Sequoias have an extensive, yet shallow, root system that spans out over an acre or more of soil. Young seedlings, sprouted from a seed the size of an oat flake, lack this extensive water support system, making them vulnerable.

For the towering Redwoods, a warmer, drier climate will have other impacts. Long ago, Redwood ancestors lived over a much broader section of the northern hemisphere but drought caused the population to shrink dramatically. Only 5% of the Redwoods that were here at the dawn of the Industrial Age are still alive today and the southern part of their range is feeling the pressure of rising temperatures and drought. Along this southern boundary, the redwoods are suffering. Picture a forest of stalwart giants dying back, their crowns thinning as they desperately calculate their next move.

This is not an excerpt from a science fiction horror show, many trees and other species are already getting squeezed into biological pressure cookers as they quickly ponder their next survival strategy. Under severe conditions of climate change, one scientist, Dr. Miguel Fernandez, predicted that up to 79% of this forest's natural range could be lost, if the trend towards hotter and drier sticks. Alternatively, if ocean upwelling brings in cooler masses of air, the trees could fair more favorably. The big question is, which way will the climate swing?

Remnant populations of Redwoods now live along a slim strip of coastline in northern California, into Oregon. Here the trees make use of the churning coastal fog to capture moisture for their crowns. Weather patterns that disrupt that fog bank could have enormous consequences. In general, moisture is a critical element for these trees at all life stages. When drought stressed, redwood seedlings lower their photosynthetic rate by 80%.

For anyone who has ever walked through this primeval forest when it is shrouded in a blanket of fog, you will intimately understand the relationship that exists between water and these towering trees. Walk for a few hours amongst these giants and your hair will dampen and a chill will settle into your marrow.

Losing a species that is so complex and mystical seems unthinkable to me. Let us hope that within the genetic code of the Redwood tree, which is ten times larger than any human's, there is an answer to this potentially pressing dilemma, for I cannot imagine the west coast of California without its damp forest of Redwoods underlain by swaying purple lupine flowers in the spring.

Because many people are worried about the status of these trees, scientists have been examining patterns of weather using 115 years of data collected nearby. Already significant changes have been discovered where the

Redwoods live. Warmer summer temperatures in the evening were the most notable, and though precipitation amounts haven't yet changed significantly, rainfall has been highly variable.

In the Sierra Nevada Mountains to the south, Giant Sequoias have been battling long periods of drought and changes to the surrounding snowpack. In cooler times, more snow fell in the mountains and a slow melt in the spring allowed for a gradual release of water. As things have warmed up, rain over snow events have caused snow melts to accelerate. This not only affects trees, who may have a longer period of dryness to combat in the summer, but it also significantly impacts our ability to store drinking water and use it for growing crops in the salad bowl of the U.S. By 2100, the Sierra snowpack is expected to drop 48 to 65% from its historical average.

Though Sequoias also suffer under droughty conditions, they do so less than their thinner cousins, at least when young. Sequoia seedlings tend to have fewer air bubbles in their water tubes and this gives them better sucking power. Still, Sequoia trees drop millions of seeds in the hopes of growing just one Giant. If warmer and drier conditions persist in California, both trees will be challenged in the future. A large Sequoia can draw up to 2,000 liters of watery sap in a single summer day. Groves of these enormous trees thus require a huge amount of water to survive.

Bitter water wars have been fought in California for hundreds of years. As the Sacramento-San Joaquin Delta dries up, it is not only trees that will experience thirst. As water levels drop dangerously low again, I would not be surprised if over-populated areas advocated for sawing groves of these Giants down, to slake their thirst. That same consciousness was once responsible for chopping these majestic trees into toothpicks, grape stakes and railroad ties.

When I sat in my Sequoia Tree on New Year's Eve, I pondered the Y2K issue...what I did not realize at the time was that there was a far more pressing problem afoot.

As our climate continues to change and resources get scarcer, unscrupulous leaders will look to the vast storehouses of natural resources found on our public lands, for supplies. If we are not careful, things we once considered sacred will end up on a chopping block where they will be auctioned, dug up, fracked, drilled, scraped away, developed, mined, burned or cut down for the sole purpose of lining pockets already bulging with cash, stripping the average American of their heritage in the process.

Twenty-three students and adults, arms fully outstretched, can circle one of these majestic trees. Cutting one down would take a few hours of hard labor. Growing the same tree back would take over three thousand years. No price tag should ever be hung on our American heritage. Public lands, and the resources found on them, are meant to be enjoyed by the masses, not doled out to a small minority, for profit.

Call me naïve, but in my vision of Making America Great Again, I see school children from all walks of life looking up at giant trees with a sense of awe, their imaginations sparked with wonder. Teachers are armed with science books, bountiful supplies and books worth reading…not guns…and Rangers and scientists are encouraged to speak the truth about environmental issues of grave concern to future generations.

OUR ACIDIC OCEANS

Mangroves, gliding pelicans and oozy mudflats characterize Florida Bay, a body of warm water that resides in Everglades National Park. Here, reddish egrets strut, flamingoes pinken themselves by gulping shrimp and gators lurk in nearby swamps. It is also where I swam with dolphins on my 28th birthday.

We had an old boat in those days and the dolphins would often play in our bow wake, jumping and sailing through the air as we made our way towards Cape Sable. On a whim on that balmy winter day, I asked my husband to stop the boat so I could jump in the water and swim with the dolphins. Intuitively, I somehow knew that frolicking with these sea mammals was going to be both safe and enchanting so I quietly slipped off the bow into the waiting hand of Mother Ocean.

Before long, the dolphins gathered. "Don't move," my husband said, astonished, "they are circled all around, looking right at you."

I treaded water with my circle of bottle-nosed friends, amazed at the communion I had with these sea creatures.

The ocean has always had a soothing quality for my soul, both in Florida Bay and at Asilomar Beach in California. Following a near fatal bounce with cancer, my husband and I spent hours listening to the waves roll onto the coast. You can wash away a lifetime of pain at the ocean's edge, sometimes in a single day.

Now our oceans are in serious trouble for like trees, our salty havens are overcompensating for our sins. As carbon dioxide levels rise in our atmosphere, the ocean does everything within its mighty power to absorb the excess CO_2 into its watery depths. The result of this action is a lengthy chemical process whose end result is an acidic ocean, with lower pH levels.

For about 250 years, since the beginning of the Industrial Revolution, CO_2 levels in the atmosphere have risen about 40%. Much of this CO_2 spewed out of smokestacks and engines, a by-product of burning billions of tons of coal. Sooty coal dust settled onto trees in England, changing the bark's hue to a darker color. Peppered moths, which were historically a whitish grey, underwent a directional color shift and began churning out greyish black moths instead.

Variety carbonara had never been seen prior to 1811 but quickly outnumbered the more typical, light-colored, pre-industrial moth. Melanism, spurred on by industrialization, enabled the dark colored moths to blend into soot-coated tree bark, keeping them safe from preying mouths, an advantage white moths did not possess. Peppered moths are a perfect example of evolution in action. By quickly adapting to shifting conditions, the carbonara moth thrived in a rapidly changing world.

If only all animals could adapt so quickly, creatures in our sea might fare far better than what scientists are currently witnessing first-hand. As the chemistry of the ocean is altered, sea creatures are scrambling to adjust.

Marine environments are shuddering due to the interface that exists between our rolling ocean waters and the spacious skies above. Like the lining of our lungs, this permeable interface enables the exchange of gases and moisture. Here the atmosphere breathes into the deep blue waters. Excess CO_2 is absorbed by the ocean, making it one of the greatest buffers we have against extreme climate change.

Due to the inter-breathing occurring between these two systems, ocean acidification is exceeding anything that has happened in the past 300 million years. Just as the ocean is sponging up the extra carbon in the atmosphere, so too are the mangrove estuaries surrounding Florida Bay. Carbon dioxide is packed away in mangrove leaves and the oozy mudflats lurking beneath the salty water.

Nature, once again, is doing its best to balance a lopsided system. By processing the extra CO_2 in this manner, the calcium carbonate that crustaceans use to form their hardened shells is silently getting eaten up in the chemical reorganization of our ocean waters. Tough shells enable sea creatures to navigate the depths safely while protecting their tender vitals from preying mouths, so displacing essential chemicals needed for shell-making is of huge concern to them.

By uptaking so much carbon from the air, our oceans are acting like a great buffer, but our sea creatures are paying the price. Already in Puget Sound, off the coast of Washington, oyster populations are plummeting. Larval oyster beds, which are impacted by acidic seawater, have been experiencing massive die-offs. Not only is this a tragedy for oysters but 85% of farmed shellfish sales in the U.S. come from aqua-culturists in the region, putting roughly $270 million dollars at stake annually along with 42,000 jobs that hard-working people rely on to put food on their tables.

If mother ocean were not absorbing this excess carbon, it is estimated that our air would already have zoomed to 450 part per metric volume of CO_2, a level that would likely cause even greater external manifestations on our planet in the form of heat waves, downpours, raging wildfires and pounding hurricanes.

Already, sea creatures we love are beginning to suffer, as the ocean sponges up the CO_2. Crabs such as Necora puber, the Velvet Swimming Crab, are starting to dissolve their shells to compensate for the increasing acidification of their body fluids. Mussels, oysters and sea urchins are also exhibiting decreased calcification rates.

A chemical called aragonite is a limiting factor for many sea creatures. In its absence, dependent creatures begin to fail. In our rapidly changing oceans, aragonite levels are dropping and Pteropods, which are an important component of the polar and subpolar regions, are feeling the pinch. Thousands of these tiny snail-like creatures are packed into a meter of seawater, forming an essential part of the ocean's food web. Zooplankton gobble them up and 60% of a pink salmon's diet is also comprised of Pteropods. Whales ingest thousands of these sea-butterflies daily, swallowing them by the gulpful.

If Pteropods can't adjust to the current lack of aragonite, there could be a massive die-off…or like many organisms right now, Pteropods could simply go on the run, hoping to find aragonite rich waters elsewhere. If suitable habitat isn't located, entire sea ecosystems could fail. Water in the northern latitudes may experience aragonite shortages as early as 2050.

The unraveling of the ocean's chemistry will be well underway by the time your children are raising families of their own, if we don't make changes now.

Oddly, as some organisms fail, others could prosper. Gangly jellyfish in the far North Sea are rapidly adapting to decreased pH levels. Certain sea-grasses and mat-forming algaes may also take advantage of the increased carbon dioxide in our ocean's waters. For each organism, a different survival tale will emerge.

Many large sea mammals will have to expend more energy to overcome super-carbonized waters just to regulate their systems. Certain fish, with pH sensitive receptors, could experience homing issues, reproductive difficulty, habitat selection problems and even an inability to avoid prey.

Though some sea creatures will bound across the evolutionary divide we are creating within the ocean, others will falter miserably. Creatures with long generational periods may not make the leap. Some species will flee poleward, hoping to outrun trouble as the ocean warms. Others will dive deep into the abyss.

Like many sea creatures that experienced a huge infusion of carbon dioxide fifty-five million years ago, scores of creatures will simply die. During the Paleocene-Eocene Thermal Maximum, ocean waters soared by 5 degrees Centigrade and became caustically acidic. Methane clathrates that bubbled up from gigantic craters on the sea floor were the likely cause of this oceanic disruption. As the methane dispersed through ocean waters it latched onto oxygen molecules floating in the sea while depriving the deep of breath. Rebalancing the system took thousands of years. Meanwhile, 30 to 50% of deep marine species became extinct.

Fisheries are already expected to falter by 2050, with drops in catch sizes predicted to hit between the 20 to 35% mark. For the billion people who rely on protein products produced by the sea, this could prove devastating. More than mollusks are at stake. Poor people all around the world rely on catching fish for food. Disrupt any part of their nebulous food chain and people will go hungry.

As you contemplate collapsing fisheries, consider that even swimming in the ocean in 2050 will be different than it is today. For starters, it will be warmer. The oceans of the world have already absorbed around 90% of the extra heat generated by trapped greenhouse gases. Without this buffering action, life on earth would be unbearably hot. This added sea heat amps up the ferocity of storms that ravage our shores.

Ocean waters have also sucked up one third of the excess CO_2 currently being pumped into the atmosphere, so as time goes by pH levels will continue to drop, increasing acidity. If we don't make changes to how we are living, the oceans could be 150% more acidic just as your grandchildren reach their "golden" years.

Your family could also readily witness 90% of the remaining coral reefs in the world disappear, entombed in aragonite-deficient seawater. Lacking the necessary building blocks to construct their finger-like structures would be problematic enough but when you couple that with over-heated water and ocean acidification, the perfect storm develops for collapsing reef ecosystems worldwide.

As coral reefs, one of the most biodiverse habitats on earth next to rainforests, become all but a memory, I will nostalgically look back on the joy I had swimming amongst vibrant coral communities off Tobacco Key, in Belize, in my youth. Your offspring may not have this same thrill or privilege.

Clown fish, parrotfish and multi-structured coral along with an octopus that turned a dusky blue before my eyes are vivid memories I will always treasure. It was pure sea-magic, something everyone should be able to enjoy in their lifetime, whether visiting Tobacco Key or the warm waters of Biscayne Bay.

The shallow estuary surrounding Biscayne Bay National Park is a very productive ecosystem that protects the northern end of the third largest stretch of coral reef in the world. Spiny lobsters, darting shrimp, waving sea fans and hundreds of species of fish cavort amongst elkhorn coral living there.

Threats to the reef include pollution, changes in the amount of sunlight filtering through the water, nutrient runoff from large-scale farming and warming seawater which can cause coral bleaching. A mere two degree rise in temperature, measured on a Celsius thermometer, could blanch the whole reef, triggering a rampant stress response.

In Biscayne Bay, this process has already begun in earnest. Elkhorn and staghorn corals have been disappearing so fast, they made their way onto the threatened species list over a decade ago. If bay waters continue to heat up, cataclysmic change could strike this National Park, devastating the reef along with a myriad of sea life.

Although "protected," climate change knows no boundaries and rangers have no means of stopping what happens outside the gates of the park. In the past 30 years, roughly 50% of the world's coral reefs have died, leaving bony white skeletons in their wake, due to bleaching events. Once the process starts it often sweeps across large patches of reef in just a few weeks. Biscayne Bay, which is hovering near its maximum temperature right now, could disappear in a matter of weeks.

How does a vibrant healthy reef turn into a coral boneyard overnight?

Corals are comprised of many colorful polyps complete with mouthparts and tentacles that sway in ocean currents. When stressed, the photosynthetic parts of the coral get flushed out, leaving ghostly white graveyards behind. If these skeletal remains are shiny white, the coral is still alive but once they get coated with stringy threads of algae, they have given up the ghost.

As a Ranger, I can't imagine saying the words, "This was once one of the most prolific reefs in the world, but now it's gone. If only we could have done something."

The mission of the National Park Service is to preserve sacred places consigned to their care, for future generations to enjoy. Climate change creates stumbling blocks for every Ranger. The only way to effect change is to share stories highlighting the science in a manner that charges people up, daring them to carefully consider the risks, not only for the parks but for the health and happiness of our families. Once engaged, everyone must take action for that is our one true, essential hope.

Americans can solve the climate change problem if we choose to…I believe this to the core of my DNA. People in other countries are already tackling it and are making huge headway. If they can do it, so can we! Americans are smart, driven, problem-solving people and when we work together, there is nothing we can't accomplish, even if our elected officials in Washington refuse to help. In our hearts, most of us know the stakes are extraordinarily high. National Parks, which are living, breathing examples of God's creation deserve better, so do the people we love.

In South Florida, reefs are endangered but so is fresh water. Snaking through the Everglades are rivers of fresh water that feed underground aquifers that supply drinking water to millions of people. As our oceans rise, intruding salt water will flood the Everglades while backing up canals, dumping non-drinkable seawater into fresh water catchments.

Engineers have rigged a series of pumps and coastal gates to keep the salt water at bay but many canals will flood if the ocean rises six more inches. Already seawater is gushing up through manholes, and streets in Miami overflow during king tides despite the fact that government officials aren't supposed to utter the words "climate change" in Florida.

Exposure to viruses and pathogenic microorganisms could also threaten drinking water as leaky sewage tanks are breached by floods, spreading germs from fecal matter randomly about town. Virulent germs churning through floodwater are not a prediction, they are a reality, scooped up into sample jars and measured by scientists.

While we are ignoring the problem, saltwater creep has already shut down municipal wells in Broward County and Deerfield Beach. Meanwhile other threatened wells are getting relocated westward, for though officials can't speak the inconvenient truth in Florida, some prudent people are taking pre-emptive measures to secure water supplies.

Just as Floridians will tirelessly search for fresh water as the oceans rise, so too will manatees, an iconic sea mammal with paddle-like flippers, a whiskered muzzle and chubby belly. These, gentle threatened sea cows flock to backwood bays in Florida, in search of warmth, fresh water and hundreds of pounds of vegetation to gnaw. As water, the most precious resource in Florida, disappears the fate of the manatee will be inescapably linked to our own.

As storm surges batter the Florida coast, eroding beaches, loggerhead turtles will also find themselves scrambling for their lives. Turtles bury their eggs on sandy beaches, many of which are narrow and easily altered by an angry sea. As young hatchlings emerge, their brains are imprinted with a visual map of their birthing beach. Even as they flip-flop their way to the sea, racing against time and hungry, pecking gulls, they feel compelled to return to the same beach as adults, inspired by the maps they carry in their head. With an ever-changing coastline this internal map and compass, which has led them home and proven true for centuries, may now be askew.

Gender also becomes an issue for turtles nesting on warm, sandy beaches for the heat of the air and sand determines the sex of baby turtles. Warmer eggs produce female turtles whereas the cooler ones hatch males. In a turtle rookery in the Pacific Ocean, researchers have discovered that green turtles are producing females at a ratio of 116 to 1 male, due to ever-warming conditions. The sex of loggerhead turtles and alligators is similarly determined by the temperature of eggs. With such imbalanced ratios, you don't need to be a scientist to wonder if these creatures will survive at all.

If there was just one story of resource devastation, we could ignore it as a fluke, but there are so many woeful tales unfolding in our parks, it is mind-boggling. The old adage seems to be true: Bad things happen when good people fail to act, so we must engage the problem whole-heartedly. Manatees, coral reefs, sea turtles and the next generation of people are counting on us.

Failure to act could prove fatal. The plight of wildlife will become our plight and their mass exodus will be followed by our own hasty retreat.

In low-lying areas around the world, including Florida, hungry displaced people will have varied responses to the climate changes thrust upon them. By 2050, the International Organization for Migration believes about two million people will be on the run from climate disasters. In places like Bangladesh, New York, Shanghai, Venice, the Maldives and Miami, there may be no high ground nearby to retreat to.

Families will abandon homes, farmlands, cultures, pictures, relics, beloved cats and dogs and all sense of hope in search of food and water. Survival instincts will grip the homeless deep in their bellies. Warlike behavior and chaos will follow roving bands of displaced people searching for basic resources. If we don't change our CO_2 emissions by 2100, that homeless number could soar to a billion scavenging people cut loose by an uncaring society. Terrorists will be the least of our worries.

Those who can't bear the thought of civil unrest, may elect to stay on their sinking plot of land, ushering a silent prayer to the grey skies above, waiting for the rising seas to wash them away to a better place, where greed and over-consumption have no rightful place. Like the Karner Blue Butterflies, people who have no inkling for war, or hope of finding another home, may simply disappear.

Though some people may go quietly, others impregnated with amped-up survival genes dating back to our cavemen days, will fight. Armed with machetes, farming tools and guns, they will hack their way to higher ground. Food and water are powerful motivating forces that have hastened more than one war to start.

It is time to examine our role in this global catastrophe for the crimes we commit against nature are about to return to us tenfold. Human suffering will be as immense as the rising seas, as disastrous as rivers of rain flooding our cities and as scorching as summer droughts withering crops desperately needed by hungry people.

Evolution of the human spirit is called for, requiring us to bring forth the better, more loving, more caring and compassionate versions of ourselves. Life on earth depends upon us making the transition to a greener, more-sustainable future. Stopping the unfathomable scenario described above from unraveling requires action, but immediacy is of the utmost importance.

Every second you spend reading this book, 300 more tons of excess carbon are sucked up by our oceans, making them that much more acidic.

Every day that passes with no action, we unleash enough heat energy into our heavens to equal the eruption of 400,000 Hiroshima style atomic bombs.

Hope for humanity and all species rests in our outstretched hands. Are we ready to accept the climate challenges thrust upon us with both action and grace?

The history, and indeed the hope of humanity are awaiting our answer.

SUPERSTORM SANDY FROM THE PERSPECTIVE OF THE STATUE OF LIBERTY

The Statue of Liberty proudly stands on Liberty Island where she watches over New York City's harbor. With torch raised, she welcomed roughly fourteen million immigrants between 1886 and 1924, a symbol of freedom and democracy to many people around the world.

At 305 feet tall, this copper plated statue has a bird's eye view of the Atlantic Ocean as well as the sprawling, bustling city of New York. From her unique perspective, Lady Liberty watched as Superstorm Sandy crashed into the eastern seaboard, surging a tidal wave of water into the streets of Lower Manhattan, covering 75% of her own island in the process while also inundating Ellis Island where so many immigrants first entered the land of the free, leaving behind poverty, persecution and war-torn states elsewhere.

Though that statue, which has stood at this harbor gazing out to sea since 1886, survived the breaking waves and howling winds unscathed, she witnessed considerable damage around her. Utilities were knocked out on her island and pathways were violently uprooted, causing the loss of 53,000 pavers. Seven thousand sixty square feet of the dock used by visitors to the National Monument were ripped away and squandered by a violent sea. When coupled with the damage to nearby Ellis Island, repairing just her immediate surroundings cost about fifty-nine million dollars.

That is not all the Statue witnessed that fateful late October, of 2012. Superstorm Sandy made landfall, so she hit the seaboard with a vengeance, pushing inland, rather than stalling out over the ocean where she would have wreaked far less damage. Instead she tossed angry waters onto shore where they flooded underground subways, submerged houses and washed away cars like straw.

From her perch, Lady Liberty watched all this mayhem and more. Eighty schools were severely damaged by the storm and gas had to be rationed in both New Jersey and New York. 7.9 million addresses, and the people living at them, went without power. Meanwhile, 171 Red Cross shelters popped up to house the homeless while 352,000 people applied for assistance to survive their losses. For though many affluent individuals were struck by the storm, thousands of people who were already living on the edge were also devastated. Sandy swallowed what little safety net they owned, in a matter of seconds.

Lady Liberty shuddered when 3 nuclear reactors unexpectedly shut down. Unlike Fukashima, which spewed endless amounts of radioactive material into the sea off of Japan for years following the tsunami, Americans on the eastern seaboard got unbelievably lucky....this time.

While this was happening, a storm surge at Battery Park hit the 13.88-foot mark. Compared to the 32-foot wall of water measured by one buoy in the harbor, this surge was meek by comparison.

Water, it turns out, was one of the deadliest parts of this storm for it whisked several people out to sea, one while walking a neighbor's dog. Flooding entombed other residents in their basements, drowning them as water rushed into lower levels of their homes, filling them with seawater. Eleven million commuters were also stranded on city streets as rising water made them impassable.

Since 1987, sea levels have notably been on the rise above Norfolk, VA according to tidal gauges strategically strung along the coast. This rise has been occurring disproportionately to other parts of the world and is expected to continue along shorelines in the northeastern part of our country.

Globally, the sea has risen 8 inches above the mark it resided at in 1900, due to rapidly melting polar ice caps. As water piles up disproportionately, Boston's ocean level could be 27 inches above normal by mid-century, if the rate of climb keeps its current pace. Norfolk, with its enormous navy base jutting out into the Atlantic, could have 2 feet of extra water to contend with…and though the military doesn't like to talk about the issue at the base, flooded streets are already causing problems. If carbon dioxide levels aren't capped soon, New York could clock an extra 20 inches of water by 2050 as well.

When hurricanes pound shores near these cities in the future, higher water levels will gush further inland, wreaking havoc on homes, utility lines, subway tunnels, sewage systems and military infrastructure vital to our safety.

In 2012, La Guardia and several other airports screeched to a halt, as did the New York Stock Exchange. Federal offices in Washington D.C. also shut down, for it wasn't just New York that took the brunt of the storm. If our statue has a sense of irony, she likely guffawed that the nation's capital slammed its doors closed, for it is here, in 2018, that Scott Pruitt was gutting the Environmental Protection Agency, the very department that could put regulations in place to protect us from such battering storms in the future, by lowering carbon dioxide emissions that are causing the heating of the planet.

If Lady Liberty could have seen beyond the New York Harbor, she would have noticed that people in states all along the northeastern part of our country suffered the consequences of Hurricane Sandy, along with several other countries. Located on the statue's copper crown are seven points which represent the seven seas and continents around the globe. For an icon that stands for hope around the world, I believe the storm probably saddened her greatly for not only did people die here, they also passed in Haiti, Canada and Cuba, making the storm an international disaster.

Within our own lands, 48 people expired in New York. New Jersey was the next hardest hit with a death toll of 12. Five people died in Connecticut and 2 more followed suit in Pennsylvania. In several other states, five additional people were struck down, bringing the U.S. death toll to 73…roughly half of all known fatalities caused by the hurricane. Suffering experienced by loved ones, on behalf of these tragedies, can never be measured nor fully understood. Some of the personal tragedies that unfolded were shared on websites, but during the Trump administration, these tales of woe disappeared from the public's eye.

Keeping people in the dark regarding the true costs of burning fossil fuels is something our Statue of Liberty would never condone, for in the land of the free and the brave, unrestricted speech is viewed as a cornerstone to democracy. She would far rather see an educated populace, where people can openly share ideas unfettered, for knowledge is power and power belongs in the hands of the people. Wisdom also enables citizens to make educated choices regarding their governance and fuel consumption choices.

When unhealthy fuel consumption leads to people dying around the world, and within the borders of her own country, this causes Lady Liberty to pause, her lamp flickering in the breeze. Quietly, she contemplated the deaths Hurricane Sandy wrought along her merciless path up the seaboard and across the Atlantic Ocean.

Aside from drowning outright or getting whisked away by the sea, some citizens expired from electrocution caused by collapsed power lines. Crushing injuries triggered by falling trees and signs killed a few more and at least one woman bled out from cuts incurred during the storm's wrath, as she tried to shut off her gas. Others fell down darkened stairwells never to return again. Family members were also killed or washed away as they loaded into cars, racing to outrun the rising water.

Along with the sea surges that washed ashore, torrential rain fell as the hurricane pounded our coastline. Over the Atlantic, 10.2 inches of rain got dumped whereas several places on land logged a seven-inch deluge. As our atmosphere heats up, due to trapped greenhouse gases bouncing radiated heat back to earth, heavy rainfalls are likely to continue for the simple reason that warm air holds more moisture than banks of colder air. Weather forecasters feel that hurricanes in the future will pump out 10 to 15% more rainfall for this very reason.

Warming ocean temperatures also spur hurricane activity. Prior to Sandy slamming into the coast, schools of cold water cod shifted northward in search of cooler water, indicating that a warming trend was already afoot in the Atlantic.

As our seas continue to absorb the excess heat generated by trapped greenhouse gases lingering around in our atmosphere, more intense hurricanes are predicted to spawn in the future. That means Lady Liberty could witness more damage on our coasts in years to come. Fewer hurricanes may occur along the northeastern U.S. in the future but those that do run afoul of this coast are more likely to be amped up storms, according to climate change scientists. Consider these category 4 and 5 hurricanes storms on steroids, for they will pack immense intensity. Partly, this could occur because the deeper warm water dips into the oceans' depths, the more fuel these low-pressure cyclones have to power them.

Another reason Sandy was so destructive is because it got boxed in by a high pressure system hovering over Greenland. This, coupled with another storm brewing on the Atlantic, cut off the storm's exit path, basically holding it in place, making it one of the top 10 costliest hurricanes in history, costing billions.

As Lady Liberty considered the enormous human suffering this one hurricane stirred up in her own harbor, she couldn't help but wonder about huge storms battering shores elsewhere around the world. She realized that even a wealthy nation staggers under the burden of a superstorm, so what becomes of people struck by hurricanes in countries far less fortunate? The very thought made her shudder for her welcoming arms were meant to gather people while providing shelter to the unfortunate.

Looking out to sea, Liberty whispered words once penned by poet Emma Lazarus,

> "Give me your tired, your poor,
> Your huddled masses yearning to be breathe free,
> The wretched refuse of your teeming shore.
> Send these, the homeless, tempest-tost to me,
> I lift my lamp beside the golden door!"

Then she pondered the fact that Puerto Rico's masses were not taken care of when Hurricane Maria devastated their island, leaving a million people without water and many others homeless. Do the words of the poem still ring true, she wondered, or are there so many climate catastrophes happening everywhere that the desire to help has dried up along with disaster funding? And yet, these storm-tossed people are citizens of the United

States. Shouldn't our own huddled masses be treated with grace when misfortune strikes, especially if we helped create the problem that brewed the storm?

Liberty considered other climate disasters she had heard about recently and realized there was a link between the water that flooded Lower Manhattan and the great ice sheets melting in the arctic. Coal dust drifting from smokestacks in China was falling onto the ice caps, along with soot from major western forest fires. Both fell onto the ice silently, unnoticed by most of humanity, but they had a major impact for they changed the albedo, or reflectivity, of the ice.

Dark, sooty colors on glaciers, and the ice shelves that hold them back from the sea, absorb more sunlight, speeding up the melt rate. Changing the ice's reflectivity amplifies sea-rise and in hurricane-struck cities, abnormally high water matters greatly….just ask the residents of Manhattan.

If all of the Greenland ice were to melt, sea levels could surge 22 feet above normal. Though it's tempting to think that's not possible, a billion tons of ice puddled into seawater here, in just four years, this century. By 2040, it is quite possible that the North Pole could be ice-free and that cruise ships will be shuttling vacationers to the far north for a glimpse of the newly opened seas.

Ice shelves cork, or hold back, rivers of frozen water at the South Pole. These flat plateaus of shelf ice, which measure up to two miles thick, hold back glaciers from dumping unshackled ice straight into the warming sea. Lady Liberty trembled, for if these slabs of ice melted the sea could rise precipitously, taking humanity by surprise.

About 14,500 years ago, within a 400 year period, sea levels shot up 65 feet due to uncorked glaciers. Florida would disappear along with a hefty portion of our seaboard were that to happen now. Our famed statue would also find her lower half gurgling in seawater.

Already in 2002, the Larsen B ice shelf collapsed and fell by huge chunks into the waiting ocean. The hasty destruction of this massive ice shelf, which stood frozen for 12,000 years, makes scientists wonder if they haven't low-balled their estimates of how fast the oceans could surge above normal. Astute scholars believe that a 9 foot rise is not out of the question by 2100. If the Thwaites glacier disintegrates, the gushing sea could rise even faster for this ice-block contains a terrifying amount of now-frozen water.

Some even theorize that the unnatural sea-rise witnessed around the eastern seaboard could be due to this rapidly melting ice, which is playing havoc with the ocean conveyor belt that shuttles sea water from one pole to the other. In this instance, however, the malfunctioning conveyor could be allowing the disproportionate pile up of water above Norfolk, Virginia. If that is true, thought

Liberty, the melting ice had a tremendous impact on the citizens struck by Sandy's wrath.

That conveyor belt, which helps to keep Europe warm in the winter, could shut down as swirling chimneys that suck in vast amounts of seawater near Greenland stop transporting water to the ocean's depths. These swirling chimneys, one of which is 6 miles across in diameter are, in some cases, disappearing or losing their whirling intensity. An excess of fresh water, which floats rather than sinks, is changing the dynamics of the conveyor belt and if it falters, water warmed by the Gulf Stream may not pour as much heat into the North Atlantic, for once the submerged waters arise from their depths along the ocean's floor, they normally drift northward, heading back towards Greenland, radiating heat along the way.

Lady Liberty marveled that excess carbon dioxide could cause so much damage, altering natural systems that have been in place for millennium. By examining ice cores and permafrost, scientists can tell that in the past 10 years, the temperature has risen a staggering 1 degree Celsius. Of the 10 hottest years recorded in history, only two happened before 1998.

They have also determined that glacial states on earth happen when the CO_2 runs at about 190 parts per million (ppm). Ice recedes during inter-glacial periods when the carbon zooms to 280 ppm. Since we have now toppled beyond the 400 ppm mark, climate changes could occur abruptly, snow-balling like an avalanche, gaining speed as we hit critical junctures, like shutting down the ocean's conveyor belt or creating drought conditions that wither the rainforests, altering the globe's climate in the process.

Cracks in the ice sheet are already pouring melting glacial water into huge gaping maws called moulins. These rivers of water further divide the ice sheet, making it vulnerable to additional thawing as they bisect, cut caverns and drill deep blue holes into the heart of the existing ice, exposing more surface area to the heat of the atmosphere. Meanwhile, calving events are chunking off huge slabs of ice into the sea, exacerbating the problem. Weather events once thought of us slow and plodding are now occurring with speed and ferocity. Liberty wondered if we were already hitting tipping points from which there would be no return.

Astonished at the changes she had witnessed since she was first installed upon her pedestal on Liberty Island, our fine statue feared for all of humanity. Change was coming fast and furious and little was being done to explain the science to the masses. Information was lacking from government websites and corporations were spending billions of dollars to hide and obscure the truth so they could further their profit-driven agendas, despite the damage it was causing all around the world.

What have we wrought, she wondered, and will my arms be broad enough to welcome all of the wretched masses that could soon land upon my shores? As a symbol of freedom and liberty, she wants us to unite as one country, to solve the problem of climate change.

The attacks of 9/11 drew us together, showcasing the immense spirit Americans can muster when faced with disaster. People of all colors and religions, economic backgrounds, party lines and ethnicities worked together to overcome. Our spirit was unstoppable.

Climate change challenges can bring out the very best we have to offer as a people…if only we pull together to confront the problem. United you stand, she thought; divided, you will all fall down.

Rise up Americans and be the people we are meant to be. We can do this! So many other countries are leading and the headway they are making is remarkable. What is holding us back?

God and all of humanity are waiting for us to choose the higher path.

THE PRECIOUSNESS OF WATER

Water that pours out of our taps is over four billion years old. It has witnessed the rise and fall of entire civilizations. At christenings and baptisms, it blesses our children while drawing families together. The Ganges, a river that flows from the Himalayas, is sacred to Hindus who believe it washes away their sins.

In nature, water enables groves of Giant Sequoias to tower majestically in the Sierra Nevada Mountains. Lake Superior, which is one of the largest bodies of fresh water in the world, forms ice-caves that rival the beauty of underground caverns, with blue-green swirls lining twisty tunnels of frozen water. Dinosaurs that roamed jungles long ago drank some of the same water that flows to our taps today.

Water is both magical and mesmerizing. Storms that roll across Great Lakes are an awe-inspiring show of force with white-capped waves crashing on shore like thunder. Float in a tropical sea and you learn to appreciate the meditative qualities of water as well.

Though pure, fresh water was once a plentiful commodity, the days of stress-free water are long gone, replaced by risky, high-stakes water that is fought over like any rare resource. Consider the water war San Francisco waged on Yosemite National Park in the 1800's in order to secure a steady supply of fresh water for the growing metropolis, from Hetch Hetchy Valley. Atlanta, Georgia recognized the value of drinking water when the city's tap water dwindled to a three-month supply in 2008. Los Angeles siphons water off of the Colorado River regularly and vied for years to gain water rights to the Great Lakes. Dreams of building a pipeline for hauling that water were eventually dashed as Michiganders joined forces to block the deal.

Taking water for granted is something we can no longer do. Pharmaceutical drugs such as antidepressants, diabetes and epilepsy meds, and traces of antibiotics are currently showing up in our drinking water. Chemicals from fracking are likewise seeping into our water supplies, adding a deadly component to water flowing to our kitchen taps.

Consider Oklahoma, the home state for Chickasaw National Monument, where fracking for oil and gas has ramped up frenetically in recent years. Fracking (hydraulic fracturing) is an extreme method for extracting natural gas and oil from deeply buried layers of shale. Drilling operators force huge volumes of pressurized water, sand and a host of toxic chemicals into wells dug thousands of feet deep in order to cause fissures in the shale layer to force out the gas and oil through cracks and pores in the rocks.

Vertical wells riddle Oklahoma, and about 19 other states, with some of them drilling down 13,000 feet into the belly of the earth. Once fossil fuel rich shale layers are encountered, operators then drill adjacent horizontal tunnels. Millions of gallons of water are forced down the vertical wells into the horizontal shoots, under extreme pressure. Chemicals and sand lace this water. Sand, a proppant, is used to buttress the fractured veins while the chemically laden water flushes out the unconventionally mined gas. Explosives are also sometimes used to crush open the bedrock, aiding the process along.

In Oklahoma, in the late 1980's, 56% of the drinking water came from groundwater which flowed to millions of residents from 12 major catchment basins fed by 150 tributary water flows. That percentage has undoubtedly risen in the past 30 years. With so many people depending upon the groundwater, fracking chemicals pose a serious threat to residents.

Once fracking water gets injected underground, it is hard to trace where it will finally come to rest due to fault lines, porous rocks and leaking wells. Locating an offending well that is polluting water can be quite difficult for these reasons. In many states, drillers are not even required to account for the chemicals they dump down fracking wells.

Over 692 unique chemical ingredients are used as additives in the fracking process. By volume, the amount of noxious fracking fluid is small compared to millions of gallons of water forced down wells. Consider, however, that on a large operation, 4 million gallons of water have been mixed with between 80 to 330 tons of chemicals, some of which are known to cause cancer. Others are so poisonous, that even in miniscule amounts, they are deadly. Benzene has the notoriety of having both properties…it is toxic at levels greater than 5 parts per billion and it is cancer causing.

The average fracking fluid uses water as its base but there have been reports of non-aqueous substances such as petroleum distillates, butane, hexane, propene and benzene added into the mix on some drilling sites, according to frac focus, a database where drillers can elect to list chemicals they are injecting down wells.

The types and uses of these chemicals are mind-staggering. Biocides are used to kill bacteria whereas acids are poured into the mix to scrub the cement walls of the wells. Crosslinkers such as borate salts and potassium hydroxide increase the thickness and stickiness of the pressurized fluids, enabling them to grab and transport proppants, which prop open fractured fissures in the shale.

Once these arteries are forced open, breakers, such as peroxydisulfate, reverse that process by reducing viscosity, enabling the freed gas to flow more smoothly through the fracked shale fissures. Plastic-like compounds called friction reducers enable fracking fluids to be pumped at a faster rate under high pressure by lessening the drag along the path.

Clay stabilizers stop wet clay from swelling which could block the arteries created by the sand props. Corrosion inhibitors prevent steel well parts from getting damaged by the acidic fracking fluids.

Pages could be filled listing all of the chemicals used in fracking while detailing their various purposes, but the point has been made. Dangerous chemicals are the lifeblood of the fracking business. When departments like the Environmental Protection Agency are gutted to loosen industry standards, we are basically writing a blank chemical check to the oil and gas industry, creating an open invitation for them to threaten our drinking water, placing it at dire risk.

If you offered me a glass of tap water in Oklahoma, I would risk thirst before swilling it down. As we frack more states, in search of what some government officials call our new "green" transition fuel, the pressing question becomes: Where can we get a safe glass of drinking water in America?

Poisoned water is not, however, the only hazard created by fracking. Hydrocarbons and noxious gases have backed up pipes, laced with enough fuel for householders to light their tap water on fire after frackers have completed have completed neighborhood drilling operations. Imagine how the value of your family home would plummet once you could never use the tap water again. The inconvenience of hauling water would be maddening.

Dirty water injected into wells also has the notoriety of causing human-induced earthquakes and Oklahoma is shaking regularly because of it. In fact, it has the highest number of human-caused earthquakes of any fracking state, out-quaking California in some years.

Six hundred thirty-nine earthquakes rumbled the state in 2016 alone, hitting 3.0 or better on the Richter Scale. Twenty-one more topped a magnitude of 4.0 that same year, and one truly jostling quake clocked a 5.8, spurring multiple lawsuits.

Even if the oil and gas industry quit forcing wastewater down wells now, earthquakes would likely continue to tremble homes through 2025 due to the fact that fracked water takes a while to slither into existing faults.

Pressurized wastewater piles up on these sensitive lines, triggering them to shudder, relieving the pent-up stress. In some cases, wavelike energy undulates through saturated rocks until it collides with a fault, causing the earth to slip at spring-loaded junctures. The deeper the wastewater is buried, the greater the potential for a powerful, damaging quake to occur.

Swarms of earthquakes can trigger the same fault line multiple times, causing a series of quakes to quiver the same area over and over again. Over 70 earthquakes have struck the Sooner State in a single week. Prior to 2009, magnitude 3.0 and greater earthquakes were rare there, averaging only two per year. Following the fracking boom, quakes of this size increased by the thousands in Oklahoma. Other fracked states are also experiencing more earthquakes in regions that were once stable. Perhaps the oil and gas industry should front residents coverage for their home, because before long, insurance companies are likely to abandon the playing field in frack-heavy states due to substantial risks.

These manmade earthquakes have shaken up residents considerably and though the governor urged people in Oklahoma to pray for oil just a few years back, homeowners may now be hoping for stable ground instead. Some Sooners have even resorted to nervously checking earthquake apps on a daily basis.

Unfortunately, not all of these quakes are small. Pawnee, Oklahoma got rocked by a 5.8 earthquake and many fear the consequences of a similar event striking Cushing. Up to 80 million barrels of oil, massive pipelines and industry infrastructure are housed at this location and a quake here would be devastating on multiple levels. Several smaller quakes have already rumbled the facility.

As if earthquakes and a glut of dangerous chemicals were not scary enough, people living close to oil and gas drilling facilities tend to be struck disproportionately with a host of health disorders, many of which impact children.

A peer reviewed Journal on Environmental Health posted a study highlighting links between fracking chemicals and impaired brain function in infants and young children. Disabilities such as ADHD, impaired cognitive function, learning difficulties and autism-related disorders were noted and advisements were made regarding the need for better regulations to be put in place to protect health.

Neurotoxicity was also listed as a side effect some people experienced when living close to fracking sites along with inflammation of the nerves, impaired psychomotor skills and respiratory impacts. A rural fracking site in Wyoming had ozone levels so high they rivaled those of Los Angeles. At 124 parts per billion (ppb), the town's ozone soared well above the EPA recommended level of 75 ppb. Methane releases can also be staggeringly high as well.

Behaviorally, people that grow up close to chemical-ridden industrial sites can exhibit impulsivity, aggression and hyperactivity. White matter in the brain can be reduced and IQ levels can sink below normal.

Premature and underweight births have also been noted along with spina bifida and depression. Though the oil and gas industry brings in jobs, there is an extremely high social cost attached to the paychecks.

A pall and stink often hangs like a shroud over these unappealing sites for they are swimming in toxic chemicals. Between 40 and 50% of these toxins can negatively affect the brain, nervous system, immunity function and cardiovascular health. Thirty-seven percent are capable of causing endocrine related problems such as advanced puberty. Cancer and mutations are fomented by another 25% of the chemical brew, according to a 2011 human and ecological risk assessment.

The bottom of the fracking line is this: the closer you live to a drilling site, the greater the risks to your health. This is NOT a green source of fuel. It is unhealthy and dangerous and because of the methane it produces, it is also extremely damaging. Methane packs a far more powerful atmospheric punch than carbon dioxide, contributing fiercely to the greenhouse gas problem warming our planet.

Nestled not far from this oil and gas fracking rests Chickasaw National Monument, a valley of peaceful, rippling waters known for its beautiful creeks, mineral waters, underground aquifers and numerous fresh springs. The Monument is unique in that it was established at the request of the Chickasaw and Choctaw Tribes who revered the healing and restorative qualities of the area's waters. Historical evidence indicates that native people were making use of the fresh springs for over 7,000 years, forging a close link between the people, the western prairie and its abundant water resource. For many of the Tribes that passed through central Oklahoma, this area was sacred.

As an influx of white settlers moved in, the Tribes became concerned that the water resources were being misused. By the 1880's, pioneers were flocking to the area to enjoy the healing powers of the water. Soon after, they began developing hotels and bathhouses to draw in more tourists, commercializing an area the Natives considered as holy.

Out of concern for what was happening, the Tribes approached the federal government, seeking protection for the waters flowing near the town of Sulpher Springs. In 1906, Platt National Park was established and the magnificent water resources in the area were entrusted to the federal government, so that all generations could enjoy the unique waterways, unimpaired. In time, the park was redesignated as Chickasaw National

Monument and entrusted into the care of the National Park Service.

Thick layers of Paleozoic rock underlie the Chickasaw area which is riddled with ancient and crumbling fault lines that snake out for miles, creating a geologically complex series of faults and folds. Chickasaw is a delicately poised and fragile system of interlinking springs, creeks and underground waterways. How groundwater flows through aquifers, springs and this intricate web of fault lines is unknown.

Predicting where injected well water will join up with underground water flows is very difficult to discover and hard to trace. Globules of oil have percolated onto the shores of Rock Creek, in the Monument. For about 200 million years, marine sediments piled up in a spreading rift in Oklahoma, buried beneath an inland sea, so it is possible that these random oil slicks, which stain boats and coat onshore objects, come from natural sources located beneath the Chickasaw area. Old worn-out oil pipelines and wells also sometimes leak. Determining where the oil is coming from is no easy task.

Risks to Chickasaw's water resources bombard the area from all directions, but the real dangers lurk outside the park's boundaries. Hazardous wastes are carted on nearby highways and illegal disposal sites crop up unexpectedly. Groundwater is pumped from nearby wells and that can dissipate water levels within the recreational area's boundaries. Packed henhouses and cattle yards also produce nitrogen and phosphorous-rich runoff that can threaten pristine water supplies as can leaking sewage lines in nearby towns. Monitoring for oil seeps and water quality are a constant concern.

High concentrations of chloride and fecal coliform have been discovered. Chloride levels were monitored closely for over 40 years and acceptable levels of the chemical were exceeded on 53 different occasions. The worst infraction was discovered in Rock Creek near an Oklahoma Gas and Electric Plant. Sodium chloride, a constituent of brine water used in fracking, was also detected but rock formations in the area are not known to be comprised of this substance....so where did it come from?

When you can't track down the source, it's easy for polluting offenders to skip town unscathed.

Under the Halliburton Loophole, the dubious Energy Policy Act put in place in 2005 by Dick Cheney, the Environmental Protection Agency was crippled. By disallowing it to regulate fracking operations, it effectively made it possible for frackers to skirt the Safe Drinking Water Act.

When will this kind of insanity stop? Water is Life. Though the Chickasaws have understood this for centuries, it is a lesson more-newly arrived Americans are learning the hard way. Consider the lead-laced water that impacted residents in Flint, Michigan, necessitating pallets of bottled water to be shipped in to slake their thirst, in 2018. Worse yet is the anchor-dinged pipeline that runs under Lake Michigan, one of the largest bodies of fresh water in the world, for the purpose of transporting viscous black tar-sand oil from Alberta, at a rate of 23 million gallons a day. In Oklahoma, class action lawsuits are starting to pour in as residents seek answers for water poisoned by what they believe to be fracking chemicals.

Water is the most precious resource we possess. It is transient, ephemeral and necessary for all life. More often than not, we take it for granted, forgetting that entire civilizations have failed when it is lacking.

When pure, water is a wonderful medicine. We bathe our children in it. After a hard day's work, we let it slide over our bodies to ease tension and wash away our trials. Without it, we die miserably in three days or less.

Can we afford to continue to hire politicians that work for the oil and gas industry under the guise of civil service? What kickbacks have they received for placing our water at risk in one state after the other? Americans are paying such a high price for their short-sightedness and greed. In the process, we may become a nation where fresh water is everywhere with not a drop to drink.

Pure, fresh drinking water should be an inalienable right. The health and well-being of American families should be a top priority for public officials. Stating that agency employees are un-American and disloyal to the flag because they are willing to stand up for these rights on behalf of us all is an outrageous and unfounded insult.

We the People recognize the preciousness of water. Protecting it is an urgent and very American matter. We just wish our politicians would realize that and stand up for our rights rather than working for the oil and gas industry while occupying space in Washington.

MYSTERY ON THE MARKAGUNT PLATEAU

Like most murder mysteries, there is an element of stealth involved with this story. Secret galleries were unveiled, clandestine visits occurred, someone being sucked dry of their resources by an enemy was rumored. Residents of the Markagunt Plateau told me afterwards that "My skin prickled," for this was no ordinary mystery. Sneaky gases called pheromones were leaked through a neighborhood wrought with tension, signaling that the area had been weakened. Once the word got out, a mass attack ensued.

After the attack, scientists searched for answers in the clues left behind. Pitch tubes and frass point towards winged, hard-shelled insects who carved their signatures on the tender part of Engelmann Spruce, leaving their calling card behind. Entomologists refer to these unique calling cards as galleries and engravings. In essence, they are tunnels dug out by the insects where they stash their most prized possession: eggs.

Spruce Beetles (Dendroctonus rufipennis) are the actors in this forest mystery and they are taking advantage of weakened trees. As trees succumb to environmental conditions, a few beetles may move into the "crime scene" hoping to lay eggs beneath a tree's bark. This low-level, or endemic, population often begins roosting in downed trees caused by drought, windfall, logging or some other disturbance.

As environmental conditions ripen on a larger scale, this endemic beetle population may begin attacking standing trees. If there are only a few attackers, healthy trees can exude enough resinous pitch to keep the bugs from getting at the trees' vital parts such as their food and water channels, or vascular system. That defensive pitch hardens creating tubes, alerting bug sleuths to the presence of an invading insect. Frass, which is sawdust minus the pitch, can indicate that the tree is so weak and devoid of moisture that it no longer has the vitality to protect itself as the bug bores its way into living tissue. Tiny sawdust piles can be seen on the bark where an insect has successfully drilled into the trunk of the tree.

In a murder mystery, it is normal to want to blame a culprit but in truth beetles are a natural agent of forest change and renewal. Peak population levels are only reached when environmental triggers hit certain tipping points. Triggers can include a decline in the health of host trees, long periods of drought, strong windstorms, trees weakened by fire and climate changes. Often these factors are intertwined. Together, they set up a dynamic system that calls for renewal and the recycling of nutrients, a task that beetles are quite good at. In fact, it is their job.

From things I've read, the National Park Service recognizes that natural processes are a vital aspect of any forest ecosystem. Insects, like fire, trigger renewal and nutrient recycling process when conditions are primed for change. Though it can be hard to witness large areas of trees die, massive beetle infestations only occur when key environmental thresholds are reached.

In nature, the downfall of one species can equal a gain for another. During beetle booms, flickers and hairy woodpeckers benefit as they drill into trees probing for beetle larvae. Standing dead snags become roosts for other birds and even the dead and decaying logs left in the wake of a large beetle kill provide habitat for mammals and other insects. Forest succession is often jump-started following large-scale kill offs as well. Aspens and subalpine fir may prosper for decades, due to less competition from spruce, especially if all available seed source for the spruce is wiped out in the process.

Spruce beetles have gotten a lot of bad publicity lately because huge outbreaks have caused widespread tree mortality across the West, including on the Markagunt Plateau where 99% of the trees examined by scientists died, with 93% of those deaths directly attributable to just the beetles. Similar outbreaks occurred across large swaths of southern Utah, Wyoming and Colorado. In Alaska alone, 3.2 million acres of forest were heavily impacted by this beetle. Billions of trees died during the 1990 outbreak, making it the largest Spruce Beetle epidemic in North American history.

But why blame the industrious beetle? Many scientists believe that these insect attacks are a direct result of forest conditions that favor these population spurts. Strong windstorms in some locals created piles of dead Engelmann Spruce, the perfect nesting material for the build-up of large beetle populations. Long-term drought can also weaken trees to the point that they can't "pitch-out" or flood invading insects with gooey sap, thus stopping them in their tracks. Research also indicates there is a strong link between increasing temperatures and spruce beetle booms.

In the next century, global temperatures are expected to rise between 1.8 and 4.0 degrees Celsius. On high elevation sites such as those found at Cedar Breaks National Monument, warming temperatures will likely create considerable ecosystem change. Many scientists are expecting more severe periods of drought in the West. This could impact the health and composition of many forest types we love, including Engelmann Spruce.

Beetle outbreaks, as well as the forest themselves, are predicted to be dramatically affected by these warming temperatures. Some insects have the ability to rapidly adapt to changing temperatures. By taking advantage of the warmer climate, they could aggressively expand their territory. It all comes down to the beetles' life cycle.

Spruce, and other beetles, often enter a period of dormancy, called diapause. This process is driven by endocrine signals and environmental cues. The purpose of this dormancy is to keep the insects in sync with the great outdoors, food availability and to enable them to survive extremes in temperature. It is also helps them to time their life cycles to what is happening around them.

How the changing climate impacts each species of beetle depends upon the life stage in which dormancy is triggered. For example, higher than normal summer temperatures could halt the pre-pupal resting period of spruce beetles. Rather than "snoozing" through part of summer, warmer temperatures can induce spruce beetles to continue to grow and thus complete their life cycle in a single year versus the two years it normally takes to create an adult insect. This amps up the ability of spruce beetles to create the next generation that much faster.

Once adult beetles emerge from beneath the bark of a tree, their mission is to find another home. Sophisticated chemical receptors on their body enable them to pinpoint a host tree. These receptors also give an indication of the tree's defense capacity. Once a weakened tree has been selected, it begins boring into the cambium, or living tissue, of the host. Here they carve galleries where the next generation of beetle eggs will be deposited.

If many suitably weakened trees are present, the beetles will emit a chemical pheromone to alert others in the area that it is time for a mass attack. This is how large outbreaks are spurred on. The beetle is simply responding to favorable environmental conditions while announcing it to fellow insects.

Though it is easy to blame the beetles for this state of affairs, to pinpoint a true culprit you would have to examine how the forest got weakened in the first place. Dig far enough into this mystery and you will soon realize that the beetles are not to blame. Their job is simply to begin the recycling process for a forest that can no longer support itself under the given circumstances. As Aristotle once said, "Nature does nothing uselessly."

In his book, "A Great Aridness: Climate Change and the Future of the American Southwest," William deBuys predicts that there will be several big losers in the climate change game and the Southwest, which is host to many of our country's most spectacular National Parks, is likely to be one of the hardest hit.

Hotter and drier weather, with greater extremes in both storms and periods of drought, are to be expected. As the climate warms and dries, Pinion pines could get squeezed out of some locations. In New Mexico, 90% of these pines have already vacated for these very reasons. Joshua Trees, which are already experiencing reproductive difficulties at Joshua Tree National Park, could try to march north only to find that there are few places of refuge in which to hide, thus greatly limiting their overall chances for survival. Outbreaks of beetles could also increase along with thirst, dusty windstorms and forest fires.

Trees may not be the only loser in this environmental game of chance we are playing with the climate. Entire civilizations have bailed from the sage covered plateaus of the southwest before, due to lack of water. In their wake, they left behind pottery shards, maize granaries, elaborate cliff dwellings and all sense of hope.

THE BIG THAW IS ON

Fire is no stranger to the Rocky Mountains, nor the hallowed grounds of Yellowstone National Park. I know, I witnessed the great fires of 1988 roll across the landscape like thunder. When the Wolf Fire jumped the road just above the Canyon junction, a jet-plane roar shook the air and trees popped as the fire arced from one side of the highway to the other, torching lodgepole pines upon contact. As a young person, I had to severely question why I was standing at the juncture that day, watching history unfold.

The year that Yellowstone burned brought many revelations. One of them was that drought-ridden trees burn with a vengeance. Another is that when conditions are ripe, no amount of manpower or money can extinguish such a conflagration on any timescale that makes humans happy. Storm force winds and burning embers ignite spot fires far ahead of firefighting resources, encouraging big fires to burn on many fronts, leap-frogging from one stand of brittle-dry trees to another, especially when ground fuel is crisp.

Stand-replacing fires in lodgepole pine are not an anomaly. These even-aged trees have burned in huge swaths in the past and will do so again. When you look at a forest of trees thicker than dog hair, growing close together and nearly the same size, there is a good chance they all seeded into the area simultaneously, following a cataclysmic event. For lodgepoles, that event is fire. Flames scorch and consume the undergrowth, opening up sunny patches, while burnt trees rain millions of seeds released from cones seared open by the flames. This is the lifestyle of lodgepole pine and newly hatched seeds rapidly colonize gaps created in the fire-ridden forest.

As our climate continues to change, Yellowstone country will likely be subjected to more huge fires. Temperatures in the park are already higher than they were half a century ago, especially in the spring. Rather than cooling off at night, evening temperatures are remaining rather warm, bouncing back heat absorbed during the day. These rising temperatures are spurring booms in pine bark beetle populations, fast-tracking their life cycle, enabling them to punch out more hungry beetles in a shorter period of time. As these voracious insects gnaw their way through stands of Whitebark Pine, they increase fuel loadings on the ground as stricken trees fall.

In the northeast of the park where Tower Falls cascade 132 feet, framed by volcanic pinnacles of rock, 80 more days a year rest above the freezing mark than just six decades ago. One could surmise that this very freezing and thawing could have pried the Tower boulder off its perch, encouraging the house-sized rock to tumble down the face of the falls one quiet day. Likewise, thirty fewer winter days cloak this part of the park in snow than they did in the 1960's. By as early as 2075, April 1st could become the new snow-free date in Yellowstone. By the end of the 21st century, temperatures in the park could be 6 to 13 degrees hotter, with the rest of the Rockies lagging not far behind.

The Big Thaw in Yellowstone is here and those warming temperatures are bound to alter the water cycle along with everything that depends upon its flow. Warmer temps and less summer moisture also ramp up fire season, extending the number of burnable days well into what firefighters once considered the shoulder season.

Fire and ice are inexorably linked in Yellowstone. The amount of snow that falls impacts the greenness of plants. If it melts off early, a longer period of drying can result wherein fine fuels crisp in the heat of July and August while small plants wither. Early melt offs could also diminish stream flows during summer months. Moose, fish and grizzlies will likely notice the alterations of this changing cycle first but ranchers, ski resorts, potato farmers and fly fishermen will also feel the differences downstream, for water is a defining element in Yellowstone.

How the water cycle will react in future years is a big unknown, probably because so many factors play a role in calculating outcomes. If the 2017 Montana Climate Assessment data can be safely extrapolated to cover Wyoming, more precipitation could be in the offing, but not all scientists agree on this element…and for Yellowstone, water is everything.

If the park had a pulse, it would be in the shape-shifting form of water: throbbing, gliding, bubbling and bursting water, coursing its way through the heart of its sage flats, yellow-stained canyons, up the xylem and phloem tubes of Whitebark Pine and percolating in stinky, sulfur-ridden mud pots.

The Lamar River meanders through the verdant Lamar Valley after coursing through rugged mountains in the Beartooth Wilderness. Grazing bison, antlered elk and howling wolves witness the river's life-giving passage through the northeastern portion of the park, where the Tower boulder fell. Further south, the Yellowstone River plunges over a precipitous canyon, cascading hundreds of feet down yellow canyon walls, billowing up spray as it drops, drops, drops on its way to Gardiner, Montana, eventually journeying to the wide Missouri.

Near a high plateau bench in Gardiner, the Boiling River steams the surrounding air, providing a backdrop to the sound of Big Horn Sheep clacking horns as they head-butt their way through the late fall rut. Meanwhile super-heated geysers, like Old Faithful, spray plumes of water onto wind-twisted evergreens lacing them with rime ice, creating a magical winter scene for the intrepid traveler to see.

So what happens if climate change suddenly alters the stream flow or tightens up the tap? Yellowstone snow feeds rivers that span our continent, slicing their way to the Pacific Coast as well sluicing down to the Gulf of Mexico. Type of precipitation, how much falls at once, and seasonality of the moisture, are all key factors, whether the amount goes up or down. Let's face it, getting slammed with 30 inches of rain in a 24 hour period similar to the April 2018 deluge in Hawaii is no better than toughing out an extended drought.

The headwaters for the Yellowstone, Snake and Green Rivers reside in or near the park. These rivers feed tributaries that pump water into the Missouri, Columbia and Colorado Rivers. If climate change significantly alters stream flows in the Yellowstone area, millions of people will feel the result downstream.

One thing is certain, changes are coming our way, faster than we could ever have imagined. Many of those changes will pack a punch, some landing squarely unexpected surprises in our midst.

Already, conditions in Yellowstone are shifting beyond their normal range. Park scientists have been monitoring snowpack levels, water temperatures, air temperature at various elevations, the frequency and size of fires and even the production of pollen, amongst other variables.

Climate changes are impacting park resources now, and the longer we wait to alter the amount of carbon dioxide emissions in the air, the greater the future consequences. What little wiggle room we have will rapidly disappear along with workable solutions if the problem continues unchecked.

Some might even argue that time is already running out. Species such as Whitebark Pine are currently feeling the squeeze.

Ancient Whitebarks, a keystone species in Yellowstone, are one of the most heavily impacted trees. As white pine blister rust spreads and more pine beetles survive warming winter conditions, huge swaths of trees are succumbing, creating a standing graveyard of timber. Parasitic beetles have gnawed their way through roughly 75% of the mature trees, in the Greater Yellowstone area, in the past 30 years and the rust, which enters the tree through tiny needle holes, erupts into oozing cankers that effectively kill branches, devastating cone production.

Just as the Big Thaw is on, so is the Big Migration. Plants, and the wildlife that depend upon them, are on the move. As the exodus unfolds, new species will reposition themselves, filling the void. By 2100, Yellowstone could look more like the southwest than the Rockies. Sagebrush and juniper will march uphill, claiming 30 to 55 percent more turf while high-elevation trees like lodgepole, subalpine fir and spruce could shrink by 50 to 85%.

Trees like Whitebark Pine, that already live in the highest, toughest subalpine zones have few uphill places they can migrate to, except perhaps some cooler northern facing slopes, where thin soils are adequate enough to form a foothold. Hopefully, Clark's Nutcrackers will efficiently disperse seeds elsewhere, allowing the tree to survive by climbing in latitude, for it would be a shame for this keystone tree to disappear as it provides food for so many species, vastly increasing the richness and biodiversity of its community.

Bears are just one of those community members that rely heavily on nuts dispersed by birds. Fat-rich pine nuts are critical for bears that have to butter on the fat to survive winter. Bears birth tiny cubs in winter dens, relying on fat reserves to survive while nursing their young. If sufficient fat reserves are not on hand, female bears can self-abort a growing fetus. Subconsciously, bears seem to know what harsh conditions they can endure and which will topple them over the edge, and their bodies act in accordance, tipping the scales on the side of survival. Reserves of winter fat help to ensure a successful pregnancy and nothing lays on the fat like pine nuts.

In years when these trees mast, producing a bumper crop of fall nuts, up to 80% of a grizzly bear's scat is comprised of Whitebark Pine nuts. In off years, bears roam elsewhere seeking other sources of food. As buffet table generalists, opportunistic bears get around…and though many tucker in when the pine nuts are masting, they all scrounge other things provided on nature's table including acorns, ants, berries, fresh greenery, fish and carrion.

Chirping red squirrels help bears to fatten up by dropping pine cones from branches located high in the tree. They then hide their cones in middens, "squirreled" away for a later date when they will chew through the cone bracts to get at the hidden seeds. Hungry bears sometimes dig up these stockpiles, robbing squirrels of their treasure. Grosbeaks also make good on the nutrient packed nuts, along with Mountain Chickadees, Hairy Woodpeckers and dozens of other species.

Clark's Nutcrackers are one of the most important birds flitting through the Whitebark Pine forest, for not only do these birds rely on the nuts for survival, eating volumes of them, but they also encourage this forest to extend its territory. One busy Nutcracker can cache up to 100,000 seeds in a single year by carting them around in a special pouch tucked beneath their tongues. Once their pouch is full, the bird finds a safe place and stashes its hoard of seeds.

Buried nuts become bird food but many forgotten seeds sprout. For a wingless seed that can't fly, emerging from a cone that can't pop open without prying teeth and beaks, this mutualistic relationship benefits both partners enormously. The fate and range of the Whitebark Pine may very well rest in the flight path and sublingual pouch of the Nutcracker.

And though the pine is worthy of saving in its own right, we often forget how linked elements are on the landscape. Whitebark Pines regulate runoff for they root in barren landscapes where other trees often decline to grow, holding fast to their desolate perches while sheltering large patches of wind-blown snow. By casting shadows onto the snow and ice, Whitebark canopies help to retain moisture, encouraging a slow melt, allowing water to trickle into montane streams later in the summer.

From the top of our multi-layered atmosphere to the depths of our great oceans and the hallowed mountains in between, overwhelming evidence exists for climate driven changes occurring on earth. Worldwide, researchers are compiling this evidence which has been mounting since the Industrial Revolution began changing the bark color of London's trees. Environmental fluctuations are piling up faster now than ever before, setting an unprecedented pace never before seen in the history of our blue globe. Along the way, it is affecting the mystical land of mudpots and hoary frost, irretrievably changing the very face of our country's oldest National Park.

Yellowstone, like all parks, has a soul and this one was born of fire and ice. Fire can be witnessed in the eye of a loping wolf and in wisps of steam rising from mystical thermal features, linked to molten heat buried long ago beneath our earthen crust.

Falling snowflakes capture the magic of ice, for the intricate geometry of an individual flake is astonishing as is the ice-clad beard of a snorting buffalo, for it shows the tenacity one has to muster to tough out a harsh winter in Hayden Valley, in deep, deep snow.

Coursing through all of this is a river and that river marks time. It has observed changes on a scale none of us can fully fathom and yet we are a witness to its wanderings. Where this river's journey will end, nobody knows. One thing is for sure: The Big Thaw in Yellowstone is on…what we choose to do with that knowledge remains to be seen.

THE MOOSE OF ISLE ROYALE

"Old Joe" was a large bull moose with one rather floppy, half-cocked ear. When I met big, snorting Joe, he was in the throes of rutting one cold, fall day at Isle Royale National Park. In love with the idea of mating, Joe was thrashing his antlers about, tearing up bushes, smacking the ground to spread the scent of his urine and raking the trunks of fir trees in a showy display of male moosehood.

Guttural grunts echoed through the air as Joe traipsed through the boreal forest visiting one female after the other while a moose researcher and I tagged along to record his amorous behaviors. Chinning the females was one of his favorite antics along with grunting and generally pissing a lot, which smelled ghastly.

Finally Old Joe decided he'd had enough of us tag-teaming him. Forcefully, he clanked around, snorted and fixed us with a look that meant absolute moose trouble. Before the researcher could yell "RUN," I was already flailing my way through the brush, racing to hit the tree line, with Old Joe hot on my boot heels. I barely cleared the trees when Joe showed up and thrashed the stand in front of me in a powerful show of force, his palmate-shaped antlers smacking one thin maple after the other, creating a loud slapping ruckus that made my heart rattle.

Though scary, his antics were part of an age-old spectacle. Joe was simply protecting his harem and the researcher and I were clearly a threat, interfering with serious moose business. Fall ruts have been a spectacle on this remote wilderness island since moose first discovered it, in the early 1900's. Since that time, this large ungulate has played a major role in shaping ecosystems of the island, largely through eating, or herbivory.

A moose munches about 60 pounds of greenery on the island in a single day. Fir needles make great fodder for an animal with a four-parted stomach. Fortunately for moose, Isle Royale resides in an area where the boreal forest from the far north dips down, meeting up with deciduous trees that prefer warmer weather.

Cold winds, rocky shorelines and chilly temperatures ensure that plenty of Balsam Fir and Spruce have graced the shores of Isle Royale in the past but as global-temperatures warm, forests in many parts of the world are shifting northward, seeking more suitable temperatures at higher latitudes.

Since the 1950's, northern boreal forest cover has shrunk on the island by nearly twenty percent. Meanwhile, deciduous trees, with their leaves held aloft in breezes blowing off the Big Lake, have increased by 15%, filling the gap. Not only can the moose not reach these airborne leaves but in the winter, when moose need calories the most, the leaves have long since dropped and are moldering rotten on the ground. Evergreens, on the other hand, provide food year-round as well as cover.

Just as the northern forest has shrunk by 20%, so too have the skulls of island moose down-sized by 16%. Skull measurements reveal a great deal the natural history of moose, including body size, physiology and environmental conditions that impact their lives.

When an animal is directly linked to a vegetation type, shifts in that plant community are bound to have an impact on the population. Even when low-lying greenery is available, some moose starve in harsh winters. Historically, wolves have kept Isle Royale moose herds in check since they loped across an ice-bridge from Canada, landing them squarely on an island paradise steeped in moose.

Weak, malnourished moose make easier prey for hunting packs of wolves. Caught in deep snow, wolves circle these moose, working in tandem to hamstring the flailing animal. Like any predator-prey battle of wits, moose do not go quietly into the dark, cold night. They fight…and very often they win, but when the prey is ripe for the fall, circling wolves will take down a 1,200 pound moose after many hours of struggle. This has been the age-old predator prey story on Isle Royale since both mammals colonized the island.

Lately, however, the balance between predator and prey has been askew. Isolated, island-bound wolves develop weakened genetic stock due to inbreeding. Malformed vertebrae in the wolves' lower backs have increased as a result. Lack of genetic variability is certainly a factor in the declining wolf pack numbers, causing them to hit rock bottom, leaving behind just two lone wolves, in 2016.

With almost no wolves left to keep the foraging moose in check, the ungulate population is exploding on the island. This has happened before and the tendency is for the unchecked animals to eat themselves out of house and home. With almost no predators and less of their preferred food base around, due to the changing climate, island moose are poised for a future crash.

Gathering baseline data is essential because in other northern states, such as Minnesota, moose have been dying by the droves as the climate warms. Granted, conditions on the mainland are different than those on

Isle Royale, but moose in both places have a serious love of one thing and that is cold.

Thick moose fur traps air near the animal, insulating them from extreme cold. Hollow hair shafts near the top layer of fur also trap air, providing a cushion against cold winter temperatures while providing some measure of buoyancy. Moose typically submerge themselves in ponds in the spring and summer searching for aquatic vegetation to supplement their raw diet. Wet moose, with aquatic vegetation dripping from their antlers, is a classic scene in the far north.

Swimming in cold bodies of water is a regular moose habit, even on Isle Royale, where they do this not only to search for food but to locate isolated islands free of wolves. These sheltered spots, which are inaccessible to hungry predators, provide safe accommodations for birthing moose calves.

In winter, moose use their tough cloven hoofs to help them walk more effectively on the snow, spreading their weight out. These same "snow" shoes enable them to dig through piles of slush and ice, unearthing buried greenery to nibble. Even their size is geared to survive the cold whereas the same bulk heat-stresses them when temperatures rise.

Though both moose herds are adapted to cold weather, isolation could be the factor currently favoring Isle Royale moose over their Minnesota counterparts.

White-tailed deer, which are not present on the island, carry a fatal brain worm they are passing onto Minnesota moose. The worm causes a host of neurological disorders including weak hindquarters and balance problems.

Though not present on the island, deer are flourishing in Minnesota. Warmer temperatures and an encroaching deciduous forest comprised of oaks and maples are displacing the boreal forest favored by moose.

Aside from brain worms, black flies and liver flukes, ticks are also turning parts of Minnesota into a moose cemetery. Warmer summer temperatures and mild winters are heat-stressing the animals while encouraging pesky insects to multiply.

Ticks are coating moose, with up to 70,000 bloodsuckers attaching to just one animal, representing a huge increase over normal insect loadings. Warmer temperatures are increasing tick numbers in the north because they provide a longer period for finding blood meals.

Moose are rubbing off huge patches of fur to rid themselves of these pests, leaving them weak and exposed in the process. Some moose have rubbed up to 90% of their fur off to purge themselves of the ticks, leaving very little furry insulation behind to protect them from chilly temperatures.

Along their southern boundary, moose populations are plummeting 12 to 19 percent each year. Further north, in Alaska and Canada, the death rate is not as pronounced, showing an upper mortality rate for moose at 12%. The southern limits of the mooses' range is apparently being tested as states along the Canadian border warm.

If climate change is the driving factor upsetting the balance in the far north, more moose mortalities can be expected in the future. How this will play out on Isle Royale is yet to be seen.

These very issues have spurred researchers to ask a number of pressing questions. What are the impacts of climate change on Isle Royale moose? How will the ecosystem be altered as species begin to shift? Will aquatic and terrestrial plant communities that the moose depend upon survive?

So many questions have no answers for we are entering a period of the great unknown. Moose, like many other iconic representations of the north woods, are in a state of flux. As climate change unfolds, many of us will be forced to let go of images we hold dear, like drifts of snow clustered on evergreen branches and rooftops. Certain bird calls, like that of the Kirtland Warbler, could disappear from jack pine forests. Even the timing of flowers and revered memories of the north we once knew could be placed at risk.

I can't help but wonder what will become of Old Joe and other moose like him on Isle Royale. The sound of moose antlers clacking the corner of my visitor's cabin is dear, flooding me with memories of chilly fall mornings on the island. Evading a rutting bull with a canoe shouldered over our heads is another image I can't shake, nor live without.

Climate change not only disrupts the lifeway of plants and animals, it alters the very connections we have formed with sacred landscapes. Somehow I feel lost in a world I once knew, where the only thing that's certain is unpredictable, and potentially devastating change.

A SACRED TRUST

Wild, untrammeled places are not a lavish indulgence, they are a necessity for the human spirit. When walking through such sacrosanct places, smelling the tang of pine and sage, we leave behind wearisome thoughts and touch the edge of the Universe. America is unique in that our forefathers had the wisdom to set such sacred places aside for the enjoyment of the people, so families could interact with and look in awe at landscapes that are sublimely beautiful to the soul.

For many years, I spent ten days each summer hiking into the far reaches of some of the most beautiful places on earth just to experience this oneness with creation. Staggering under the weight of a heavy backpack, I hiked up 4-day mountains for the sheer joy of feeling like an ant, kicking into knee high snow, struggling up the side of an ice-clad mountain not only for the view but so I could recognize my true place in an infinite Universe.

Sometimes feeling small is the most rewarding feeling of all, for it makes you appreciate vastness. When you can lift your arms up and touch the sky you know your prayers will be instantly heard for you are already perched on the edge of Heaven.

My entire life, I have been working under the assumption that as a nation, we would value America's Best Idea, our National Parks, to the point that we would do everything needed to protect them. The peace, the sanctity, the wildness and the overwhelming feeling of immensity they offer seem like reward enough.

These dynamic landscapes are also home to a vast variety of plant life, flitting birds and intriguing mammals, a place where ecological processes are given vital space to function and unfold on a somewhat natural level. Meanwhile they provide respite from urban jungles and the frantic pace of our daily lives, allowing people to re-create themselves on trails, bike paths and in canoes paddling down wild and scenic rivers.

With the potential repeal of the National Parks 9B regulations, I now know this assumption is false. Under the Trump administration, agency heads are being directed to review all current regulations that could slow down the development of oil and gas drilling in our country, including on lands entrusted to the care of the National Park Service. Prior to leaving Washington, Secretary Zinke had been empowered to suspend, revise or remove the 2016 9B rules if he saw fit, as a result of the executive order Trump signed in 2017.

This same administration wants to promote offshore drilling in 90% of our U.S. waters, so why should any of us be surprised? And, as if the BP oil spill were but a horrific distant memory, recent tax cuts put in place by Congress basically unfunded the Oil Spill Liability Trust Fund, meaning that if companies drill offshore and a spill occurs, it will be left up to affected communities and states to clean up the oily mess. This sanctioning of corporate polluters is yet another sign showing how far off-track our government has gotten. Rather than working for "We the People," they are obviously functioning on behalf of "Thou the Gas and Oil Companies."

This complete disregard for our ocean's waters, public lands and parks makes me want to weep. Not only is our current administration setting a horrible precedent, but it makes me fearful over what additional tragedies will unfurl as we hire other non-ecologically aware staff members to direct agencies meant to protect us from environmental wrongdoings. If we need trees, will we start logging our Parks? Already we have rescinded protection from Bears Ears National Monument so it could be opened up to mining uranium nearby.

Meanwhile, Congress has even suggested turning over huge tracts of federally owned land to the states to help balance the budget. Once received, the question becomes, who will manage these lands for the people? What will they be used for…casinos, fracking, property development or some other atrocity? Never have I seen such an unenlightened view coupled with complete lack of understanding for what our parks and public lands stand for and I fear that the pillaging of America's most beautiful places has begun in earnest. Along the way, I wonder if any thought was given to the fact that these lands are held in trust for the American people.

According to Yessenia Funes at earther.com (https://earther.com/trump-suggests-we-just-drill-everywhere-1821781890) a draft version of our "great" American budget, that was leaked, contains a proposal to route pipeline projects affecting National Park Service lands through the Secretary of the Interior alone, rather than engaging Congress. If this pans out to be true, let us hope that chairperson will shoulder the sacred trust we have placed in him to protect our parks, even if he was formerly a lobbyist for the gas and oil industry. If the leader for our public lands is not up for the task little will stand between our parks and their ultimate destruction, unless of course, America speaks up.

One-hundred and nine incidents involving the movement of gas and oil cost $33.5 million dollars last year, according to statistics recorded by the Pipeline and Hazardous Materials Safety Administration, so the chance of having a huge spill on sacred ground runs very high.

Because of the inherent risks of drilling in fragile ecosystems, the National Park Service established the 9B rules in 1978. Then, as they discovered loopholes and weaknesses in their own regulations, the agency worked on them over a period of 7 years, while engaging a variety of constituents including the American public.

The intent of the 9B regulations is to limit drilling in National Parks and to improve operating procedures in the dozen parks where companies are already drilling. The NPS realized they needed to refine the rules because 30 additional units, including Gulf Islands, Everglades, Grand Teton, Mesa Verde and the Flight 93 Memorial have what is considered split real estate wherein a shrinking workforce of flat-hatted Rangers are to protect all that is above ground, while other entities have rights to that which lies below the surface.

The 9B revisions, which were crafted after much debate and deliberation, were put in place to minimize damage and to ensure that drilling operators cleaned and closed wells upon completion of their project. A small cap fund of $200,000 was set up so companies could be held responsible for closing abandoned wells and for restoring the land's surface to its prior pristine state. That amount, which was established in 1978, is a pittance compared to what it would actually cost to rehabilitate drilled sites to a natural state today.

By writing the revisions into law in 2016, the NPS was helping to ensure that taxpayers would not be tapped to correct deficiencies left by the oil and gas companies which could run as high as $12 million dollars. Paying the cleanup costs for private industry doesn't seem like something the average taxpayer should have to shoulder.

The revisions also address roughly 60% of the wells already found in the parks that were not even covered by the 1978 regulations. By repealing the 2016 law, parks that house these wells will be left powerless to address industry accidents such as leaks and spills.

Until now, this has basically been a non-issue because most politicians recognize the sacredness of National Parks, and other public lands, to the American people. Paul Gosar, however, is not one of those politicians. Under House Joint Resolution 46, the 2016 updates to the 9B Regulations could come under attack. If Gosar's resolution passes, not only will the 9B rules that affect the parks be repealed but it will prohibit the agency from ever changing them again, leaving parks vulnerable.

If you love any of these places, https://www.npca.org/resources/3190-national-parks-affected-by-9b-rules, it is time to speak up. If you would like to see better oversight on the 12 NPS units already being drilled by the oil and gas industry (https://www.nps.gov/subjects/energyminerals/upload/nps48_recommendations_20140507_FINAL.pdf) please talk to your representatives in Congress today. The intent of these laws is to help protect the parks for generations of families yet to come, while potentially saving taxpayers money.

As of October, 2017, Zinke reported on policies that could be holding back the gas and oil industry but so far, the NPS 9B rules have not been mentioned. Thank you Mr. Zinke for upholding your sacred trust while you were in office and let's hope your replacement is equally up to the task. Also please recognize that employees who are concerned about such matters are not disloyal to our country nor the American flag, they simply want to preserve the sanctity of our parks and public lands on behalf of the American people. Oil and gas pipelines do not belong in our parks, it is a blasphemy, especially when you consider the threats climate change poses to our nation and humanity at large.

For anyone reading this book, I ask that you offer up some kind thoughts or a prayer, in your own fashion and faith, on behalf of our parks, public lands and our country as a whole. There has never been a point in time when it has been needed more for we have entered a dangerous era in the history of our nation.

People who sign petitions are viewed suspiciously though many are simply concerned about the severe and unbridled damage climate change is inflicting on humanity and God's wondrous creation. Chaplains, who offer up prayers for the poor, are castigated and outed from their positions. Native people fighting for clean water have had trained dogs attack them and have even been gassed while peacefully protesting. When Pope Francis elected to speak the truth about climate change on behalf of all people, members of our government spurned him. That same government is openly supporting a war on science by unfunding projects aimed at finding the truth, while publicly denying and clouding the waters on climate change.

One must ask the question why? Why would any politician place our people at such grave risk?

Whether you are sitting on a mountain touching Heaven or thinking kind thoughts in church, or even sitting on your back porch contemplating the future of our planet, the time for change is upon us. Catalyzing that change is an urgent necessity, for our lives depend upon it.

Speak to the hearts and minds of our politicians and encourage them to consider all that stands to suffer before it is too late. When they were asked to watch over our National Parks and other public lands, they were given a sacred trust by the American people. I don't know about you, but I want them to uphold that trust for what we do to nature, we do unto ourselves. I also want them to do everything they can to halt our changing climate. Building more pipelines, no matter the location, is simply not the answer.

We are on a crash-course to disaster and our politicians are sound asleep and snoring at the wheel. Some may even be profiting from promoting the gas and oil industry at everyone's expense. Wake these people up and hold them accountable.

We have no Planet B.

UNSTABLE GROUND

Certainty, stability, the known…these are some of the things Americans seek in life. So what happens when the very ground you stand upon becomes unreliable, squishy, and subject to uncontrollable change governed by ever-rising temperatures? As an arctic state, where climate change is happening twice as fast as the lower 48, many Alaskans are learning what it means to live life on unstable ground.

Alaska is a huge state, accounting for roughly 20% of the landmass of the United States. Barrier islands where bowhead whales glide, singing soulful melodies that help them navigate through the ice, along with frozen tundra sporting ancient looking musk ox, coincide. Enormous swaths of boreal forest, comprised of black spruce, trembling aspen and red cedar, are splashed across hillsides. Tucked amongst these tree havens are peat lands packed with decomposing plant material and meadows. Lush temperate rainforests dripping with plants are an unexpected, yet diverse, part of Alaska scenery as well.

Hosted amongst these varied ecosystems are 16 National Wildlife Refuges. Sixty percent of the state's land base also falls under the care of National Park Service. The "Last Frontier" state resides on a pivotal edge, where the effects of climate change are extremely potent for along this once-frozen border the average overall temperature has already jumped 3 degrees Fahrenheit in the past six decades, with temperature increases soaring double that in winter.

As mercury jumps in thermometers, fields of ice shrink, impacting the patterns of native people accustomed to living, hunting and maneuvering on ice. Ice sheets break up far earlier in the spring and large-scale fires rip their way through black spruce forests requiring massive mobilizations of fire-fighting resources annually. Warmer temperatures have also enabled voracious beetles to go on the attack, chewing their way through 450,000 acres, killing millions of trees in 2017. Mass attacks of this nature are no stranger to the far north; in the 1990's beetles ate their way through enormous stretches of pristine Alaskan forest as well.

For the 229 federally recognized tribes of Native People living in Alaska, these sudden changes are disruptive. Additionally, game they rely on for sustenance is rushing to adjust or getting displaced.

Bowhead whales, which frequent ocean waters around Barrow, Alaska could come into conflict with shipping lanes as ice melts away, opening up a profitable corridor for vessels hoping to avoid the Panama Canal. At the same time, thinning ice shelves are prohibiting native Alaskans from venturing onto the ice to hunt whales circling nearby. Seals that perch on these same frozen shelves have discovered that 40% of their ice disappeared in a 29-year period ending in 2007, with even more ice puddling up in the past decade.

Fires, which already rage through the state, are predicted to double in size by 2050 and though Eskimos have over 100 words to describe the qualities of ice, they have only one for wildfire: ikkuma. Despite the lack of descriptive words to explain current conditions, these fires scorch rock-clinging lichens, robbing caribou of their primary winter food source and Eskimos of meat.

As temperatures soar, ground that was once frozen is now melting, causing landslides and coastal erosion to occur, slumping native homes off eroding hillsides into the waiting sea. As of 2003, a government accounting office report identified 31 villages facing imminent threats from Alaska's rapidly changing climate and yet little has been done to help villagers resettle.

As if these social and environmental problems were not disconcerting enough, the largest pool of mercury on earth is about to be unleashed from melting permafrost soils. Nearly ten times more mercury than we have pumped out of coal-burning smoke stacks, and from other polluting sources, in the near past, could wend its way up the food chain as the permafrost leaks toxins into the environment.

Scientists discovered the toxic stockpile, which contains more than 32 million gallons of mercury, when they drilled soil cores on thirteen sites in Alaska. Once the leaked mercury enters waterways, microorganisms can transmute the toxin into the more dangerous chemical variety methylmercury, known for causing birth defects and a host of neurological problems in humans, including "Mad Hatters" disease. Worse yet, methlymercury travels, so impacts will be felt thousands of miles away.

Methane releases will occur simultaneously for frozen arctic soils also contain tons of organic material buried long ago. As these frozen plants thaw and decompose, tremendous amounts of this potent greenhouse gas could leak into the atmosphere from deep pockets of melting soil and hidden fault lines. Microbes nibble away at the carbon-rich plant material as it thaws. In the presence of oxygen, the byproduct of their buffet is carbon dioxide but in an anaerobic environment, methane is outgassed instead. Either way, both greenhouse gasses heat the atmosphere, warming our planet in the process.

Although 1,400 gigatons of this plant-linked carbon is buried in the permafrost, it is our daily CO_2 emissions that concern scientists the most. It is also the only source we have much hope of directly controlling. While we contemplate the variety of ways the climate is being altered, the once solid earth in frozen Alaska is becoming squishy.

When the earth slogs under your feet, critical infrastructure is put at risk. Consider the impacts of melting permafrost on highways, homes, oil pipelines, railroads, schools, airports and even dog sled runs. Also, as heavy layers of ice melt from the Alaskan landscape, the earth is rebounding, bouncing back from shedding its weighty, frozen skin. Like the soil itself, life in the far north is at a mutable point in history and an entire culture is being put at risk.

With an uncertain, yet melting future before them, Eskimos are also in the process of losing their past. Storms are whisking away historical records at a break-neck pace, leaving archaeologists scrambling to preserve the past along the Alaskan coast. Baleen whale tools, styles of clay pottery and detailed Ipiutak wooden structures tell the stories of Native People in Alaska for thousands of years. An eroding shoreline is enabling storm waters to eat away archeology sites perched along the coast, unearthing long buried bones from ancient graveyards, washing them into the watery abyss.

Alaskans have always lived life on the edge. They are accustomed to frigid cold and ice. In a place where you can spit and have the saliva freeze before it hits tundra, blubber is an idolized gift offered up by whales and seals at the end of a successful, yet bone-chilling hunt. Watching over these people are shimmering green lights that dance across the night sky, connecting the Alaskan wilderness to the Heavens.

May these northern lights cast the perfect hue for finding a way forward, for nowhere is climate change unraveling as fast as it is in the Arctic where so many National Parks and native people reside. In a place where residents are connected to the daily rhythms of land and animals, life is becoming very unsettling in the far north, in more ways than one. Impacts they feel will sadly become part of the historical story that must be told, for here the people and the land are one entity, each impacting the other.

As we witness how things unfold in the last frontier, bear in mind that the fate of native Alaskans could easily mirror our own destiny. Toxins released in the arctic will journey their way across the globe, bioaccumulating in unsuspecting seafood. Rampant fires will chew up bug-infested timber, leaving millions of acres scorched. Homes lost to eroding shorelines and melting permafrost could set people wandering once again.

The past is melting and crumbling away and the travails of the future have been set in motion. Hidden in the vast arctic, I wonder how many of us will miss witnessing the potential decimation of an entire culture? There are no firm answers in climate change, especially in Alaska, where life is literally resting on unstable ground.

Meanwhile as the arctic melts, countries are anxiously circling the north pole in ice-breakers, hoping to plant flags and lay claims under an international treaty that enables nations to assert domain on the floors of the sea, claiming rights to the watery depths as part of their continental shelf territory. The prize for asserting domain is that roughly 22 percent of the world's undiscovered oil and gas reserves lie buried beneath these ocean waters.

Oil companies, and the politicians that flagrantly back them, will create hell on earth to extract this oil, for that is the sickness of fossil fuels. We always want more, regardless of the human cost. Our only hope is for humanity to find a way not to need those fuels gurgling beneath the arctic ice.

FIRESTORM

The Coming Storm

"Some say the earth will end in fire, some say twill end in ice. From what I've tasted of desire, I hold with those who favor fire."

These words, penned by poet Robert Frost long ago, could not be more appropriate for the coming firestorm America is facing. Already, across the mountainous west, Alaska, the sawgrass prairies and pinelands of Florida and the arid southwest sage flats, fires are getting bigger, hotter, more intense and dramatically devastating by the decade.

Fueling this firestorm is a variety of factors, including climate change. Not only is it driving long-term changes in forest structure but it is also impacting the health of certain trees, while stimulating population spurts amongst voracious wood-gnawing beetles.

As carbon dioxide heats our atmosphere, temperatures climb. In many areas, these warmer temperatures also drop the humidity in the air, creating parched conditions that sap moisture from fine fuels, making them brittle dry, raising what firefighters call the PIG, or probability of ignition. In some locations, such as fire-prone California, those dry conditions have stretched into serious droughts. There, catastrophic fires are inexorably bound to water. In 2014, all 18 of the National Forests in California were withering in drought, which put the health and flammability of many watersheds at great risk.

Parched, dense forests burn with extreme intensity leaving watersheds exposed to the risk of sedimentation, debris flows, rockslides and even flash flooding. On a hot August day in 2013, the Rim Fire consumed nearly 90,000 acres in massive fire runs spanning over two days, leaving a ghostly graveyard of charred sticks behind. This nuclear-like explosion created large gaps in the forest canopy, resulting in very high tree mortality and acres of exposed burnt soils.

Under such parched conditions, additional live trees succumb to drought stress, making them easy prey for Pine Beetle infestations. Beetles can sense weakness in trees and when an area is strained, they attack, often in massive numbers. Once devastated by gnawing insects, thousands of acres of dead and dying trees add to the fuel load, aggravating fire conditions. With no break in the drought cycle on California's horizon, future catastrophic fires posed a real threat to surrounding watersheds which supply tap water to thousands of people living downstream.

Complicating this is the fact that earlier snow melt dates are now jumpstarting fire season sooner each year, stretching it out. In the fall, late seasonal rains that fail to show up are also taking a toll. Firefighters have long had a statistically based saying in the Sierras: There is a 90% chance of getting a fire season ending rainstorm by the third week of October… but with our changing climate, those fall rains seem to get held off longer and longer. Between the earlier melt dates, warmer temperatures and late season rains, fire season now lasts about 2.5 months longer than in the historical past, yet one more repercussion of our changing climate.

The infamous Camp Fire, which blasted through the town of Paradise, California, destroying 13,972 residences, 528 commercial buildings and killing 85 people, is a perfect example of the increasing phenomenon of late season firestorms. Ignited on November 8th, 2018, the Camp Fire burned into Thanksgiving weekend, leaving 153,336 acres of torched land in its wake, along with hundreds of unaccounted for residents. This fire followed on the heels of the hottest July on record for the state of California, leaving even large fuels desert-dry and receptive to the flaming front.

Since the year 2000, more than half of western states have had their biggest fire conflagrations ever, but the damage is not limited to the intermountain west, nor California and Alaska. The Great Plains are now burning too. Between 1984 and 2014, four hundred percent more acres have been consumed by fire on Midwest grasslands and temperate forests.

Overall, between 2000 and 2016, wildfires have torched at least 3.7 million acres annually, with the exception of just three years. Since global warming has ramped up, 16,400 more square miles of fire have burned in a 30 year period than would have occurred had greenhouse gases not heated our environment according to www.climatesignals.com/climate-signals/wildfire-risk-increase.

NIFC, the National Interagency Fire Center, tabulates fire data for the country. They discovered that of the 10 years with the most scorched acreage in the U.S., nine have occurred since the year 2000, closely aligning megafire years with some of our most sweltering summers on record. Worse yet, by 2050, these roasting conditions are predicted to double the area burned by wildfires according to satellite data and computer models.

In light of this it seems unfair to state that our changing climate is not impacting the ferocity of our wildfire problem, but fire is a complicated beast and a number of factors are at play in the megafire trend we now see spreading like a red stain across our heartland. Because the problem is complex, breaking the cycle of catastrophic fire is going to take a multi-pronged effort that includes restoration, public education and thinking outside the box, especially when it comes to policy and funding. Politicians also need to take the time to educate themselves on the complexities of fire.

Changing Perceptions on Fire

Changing the public's perception of fire, through educational efforts, is a step that must be undertaken if we are to unravel the growing threat to our communities. Humans have an innate fear of fire and for good reason. It threatens our homes, blackens landscapes we love and creates havoc with the infrastructures that keep our society humming along. Searing fires melt electrical poles, dump tons of ash into reservoirs, create landslides that flow over homes and block roadways with downed trees that fall like jackstraw following a firestorm. Smoke is an inconvenience and potential health risk that keeps residents dialing politicians in an effort to shut down fires quickly, to avoid the nuisance of unclean air.

Though fire poses all of these risks and more, we have forgotten along the way that they are also a natural part of many ecosystems. Keeping them out literally adds more fuel to the next fire, for every fuel type has a natural fire cycle or return interval. Miss enough cycles and the fuel piles up to cataclysmic levels. Missed fires also change the very structure of the forest over time, making them more flammable than ever.

In response to humanity's desire to rid itself of pesky smoke and wildland fires, the U.S. Forest Service developed the 10 a.m. rule. Under this firefighting strategy, fires were to be stomped out early, keeping them small. By jumping on fires with such ferocity, firefighters successfully managed to squelch the vast majority of blazes within a day of their discovery.

As equipment, access and the skill level of trained personnel improved, fire crews became efficient fire killing machines, halting fires before they had a chance to spread across the landscape. Between 98 and 99% of fires were doused early, covered with dirt or otherwise boxed in before they had a chance to grow.

Unfortunately, several problems are linked to this fire-extinguishing efficiency. Fires have a place on our landscape. One of the biggest benefits they provide humanity is the cyclic removal of dead debris from the forest floor, including pine needles, layers of dried duff, small parched twigs, low-hanging tree branches and even young shrubs and trees. By cleaning up this burnable material periodically, the structure of the forest is kept more open, sparse and park-like. Small, encroaching trees are removed, creating sunny gaps in the forest.

In the absence of periodic fires, trees begin to crowd one another, massing together while they fight for space and light. Sun becomes a limiting factor and as the denseness of the forest thickens, shade loving trees, which were absent before, start creeping in, As they grow up and through the taller trees, their branches intermingle, creating connectivity not only with the fuel on the ground but the vertical fuel bed housed in the trees themselves.

If the forest burns regularly, these ladders of fuel are not continuous enough to allow the fire to spread from the forest floor into the tops of the trees. Once fuels become thick and touching, the fire can climb its way into treetops where it spurs crown fires. Racing crown fires allow flames to jump from treetop to treetop in a blaze that stretches several hundred feet tall while roaring like a freight train. If aided by steep terrain and upslope winds, these all-consuming fires can wipe out vast swaths of forest in a matter of minutes, creating conditions that no firefighter can outrun or safely fight.

Nineteen firefighters met such conditions on the Yarnell Fire, in 2013, and they paid the ultimate price in a boxed-in canyon, not far from their safety zone. Unable to reach their intended destination, they hid in tin-foil like shelters while the blaze roared overhead. The fire was so intense that their body bags bore numbers instead of their names because they were burned beyond recognition. Climate change has altered the rules of the firefighting game, making it far deadlier than ever to combat wildland fires on behalf of the American public.

The intensity of a running crown fire is so blistering, it can melt aluminum let alone human flesh. Socioeconomically, megafires are also very damaging. The Rim Fire, for example, impacted the power and water supply for 2.5 million people in San Francisco. Nine hundred thousand dollars were spent in order to purchase alternative energy for those impacted. Meanwhile, Highway 120, a busy corridor leading up to Yosemite National Park, was closed which impacted local businesses as well as visiting tourists.

Residents were chased out of 5,000 homes and eleven of them burned to the ground, leaving people with nothing but what they had hastily thrown into cars on their way out of town. School children were sent home due to health issues caused by smoke and several people working in the burned area were killed.

The local economy lost approximately 1.8 billion dollars and suppression efforts soared to 127 million. Rehabilitating the area after the fact cost roughly 10 million more but due to the high risk of mudslides, falling snags and erosion, the Burned Area Emergency Response Team's help was deemed an essential part of the recovery effort.

In the absence of periodic fires, damage caused by such firestorms are immense and battling them is now no easy task. For starters, we have created conditions wherein firefighters are no longer effective. Because we have effectively stomped out so many fires at a small size, the blazes now being confronted have the upper hand with extraordinary levels of fuel piled up on the ground just waiting for a spark. Additionally, 44 million homes are now scattered in the wildland urban interface (WUI), the zone where thick forests abut private property and homes.

Though some of this WUI is located in densely populated areas like southern California, many of these WUI homes are tucked in rural locations, surrounded by flammable fuels with poor road access. If homeowners are not creating defensible space around their property and using wise building materials, their lives and property are now subject to the whims of catastrophic fire. In a report created by Headwaters Economics, seven times more homes burn in wildfires than they did just 45 years ago. In 2015 alone, over 10 million acres were consumed by forest infernos, and as our climate continues to heat up, that burnable acreage will increase exponentially.

Management practices do need to change, more on some lands than others, but in order for that to happen society must first learn to understand and live with some degree of fire, then budgets need to be adjusted to accommodate the tactical switch. The danger that exists now is that so much excess fuel has accumulated making it harder and less safe to switch strategies.

Prescribed fire is one solution but it must be funded on a level that can effect positive change. And though prescribed burns are planned fires wherein fire personnel have a far greater degree of control than in an unplanned wildfire event, risk factors still abound. Liability for losing such burns scares off many well-trained firefighters who know that fire must be returned to the landscape, but are concerned about the personal risk. Likewise, local people who feel threatened by fire must learn to live with some level of hazard, fire and smoke in order to hold off greater conflagrations.

Since 1970, 98% of wildfires have been stomped out at 100 acres or less. The remaining 2% are the fires you watch on CNN such as the Yarnell Fire, South Canyon, the Camp Fire and High Park. And though we would love to continue to keep fire out, it is a very short term answer to a long term problem which acerbates the threat for all homeowners living in the WUI. Climate change is only going to make that problem far worse.

When conditions such as record temperatures, roasting droughts, steep terrain, high mid-flame winds and low fuel moisture levels align with an unstable atmosphere, conditions have been set for another perfect storm to rage across our wildlands. Under perfect storm conditions, intense heat from the ground fire forces hot air into the atmosphere. If that atmosphere is unstable, it has an upward pull or draw on the fire much like a chimney flue draws heat and flames up the pipe. From this sucking action, an enormous pyrotechnic cloud forms drawing up hot embers, smoke and ash, sometimes to a height of 40,000 feet.

Under extreme conditions it can even rip sage brush out of the ground, creating a whirling tornado of fire that flings burning bushes about its circling vortex. If this column of smoke and ash gets sheared by high winds aloft, breaking the cloud's column, hot burning embers can rain down a mile away from the head of the storm, allowing the blaze to leapfrog from the flaming front into unburned fuels far ahead.

Firefighters can only be so effective at keeping a fire genie of this type in the proverbial bottle. Firestorms like this necessitate that crews pull back and attack from a great distance, while they strategically box the fire in using distant topographical features to their advantage. Though far safer, the downside of this tactic is that the fire explodes in size. Meanwhile, towering pyroclouds spew smoke and ash across the country, landing soot as far away as the Arctic icecaps.

Unfortunately, simply raking the forest floor will not stop these conflagrations either. Fuel types are far too complex and there is not enough manpower to undertake even more reasonable fuel treatment options like creating fuel breaks around fire-prone communities which requires removing trees, layers of duff, small bushes and parched needles from within close proximity of endangered homes. Completing such projects also requires funding and manpower and it's not a once-off; fuel breaks must be maintained on a cyclic basis.

Fire Ecology

Aside from ridding forests of excess fuel, wildfires are also an important force shaping the ecology of our public lands. Ecology is the interaction between living organisms and their environment and though we often like to separate ourselves from that story, humans are key players in that unwinding saga. Not only do we witness priceless occurrences in the outdoors, but we also act as great agents of change in many forest types. Much of that has to do with how we choose to live with fire…or not, for many vegetation types require fire for survival.

The California landscape provides one example of ecology in action but fire tolerant cohorts of plants live throughout the west and many plants rely on fire as part of their survival strategy.

Huge wildfires can leave behind a devastated landscape of burnt twigs. Charred areas can show signs of life but it may take many years before blackened fields return to their pre-fire state. Though fires can be destructive, they also excel at recycling nutrients. Plant life often flourishes following a forest fire for there is a huge nutrient pulse that energizes the soil. Other plants actually have seeds that need fire to crack their seek coats so they can sprout.

Fields of lupine are one species that readily colonize burned areas in California. Gilia, Indian Paintbrush and morel mushrooms spring up as well, using newly opened habitats created by fire. From experience I know that morels flourish following a fire, for I have seen them in abundance, sprouting beneath Sequoia trees while taking advantage of the new soil substrate created by wildfire. Nutrients from dying trees infuse the ash and the mushrooms work diligently to break down the decaying matter. Though fire is almost always viewed as destructive, that is only half the story, for it also kickstarts ecosystems into starting over.

Fire enriched ash also causes growth spurts amongst native Cacao plants and Mountain Misery while feeding underground bulbs belonging to Soap Root plants. Flush with nutrients, these bulbs, which survived the fire underground, respond quickly, shooting up large strap-like leaves once the flames have passed.

Golden Ear Drop seeds lurk dormant in the seed bank for many years waiting for a fire. When charate and other nutrients are released by the passing flames, the plant is triggered into blooming. Tall stands of this unique Bleeding Heart were visible in the foothills impacted by the Rim Fire within two years, their yellow bobbing flowers swaying in the breeze. Scorching causes a similar reaction in other species, increasing germination in fire tolerant species. The message is very clear: When under fiery conditions, plants must either reproduce or die. Those that grow up in a fire prone area know the rules of survival.

Dense Chaparral thickets are also designed to burn. Volatile oils found in chamise, promote the rapid and explosive spread of wildfire. Though chamise and shrubs such as whitethorn are top-killed in the fire, many of them simply resprout from their base once the flames have passed. Oaks deploy a similar method, sprouting from root crowns, sending up ringlets of tiny oak leaves from the bases of their burnt stumps.

Extreme lack of water plagues many California brush fields, including chamise. With plants tightly spaced, all searching for scanty water, these hillsides are born to burn. The miracle is that they survive the dry conditions so well in between the passage of the flames. Many survive these droughty conditions by entering into a stuporous state of shutdown to conserve much needed water.

Thick, waxy leaves retain precious water in some species. In others, breathing holes are located on the undersides of leaves where they are safely protected from drying winds and sun-blast, thus limiting the loss of water through evaporation. Manzanita plants even orient their leaves away from the sun to lessen the loss of vital moisture. Buckeye trees go so far as to drop all of their leaves in the heat of summer to avoid desiccation by drought.

Higher up in the mountains, serotinous cones help trees such as Lodgepole Pine survive wildland fires. These cones, which are glued shut with resinous sap, need intense heat to open, allowing them to rain thousands of seeds onto the newly scorched forest floor. Stand replacing fires often zoom through thickets of Lodgepole but the gaps created in the forest create rich nutrients and necessary sunlight to spur on the new growth.

In the Sierras, Giant Sequoia trees are also well adapted to fire. Thick layers of reddish bark insulate the trees from passing flames. Sometimes, especially on the upslope aspect of the tree, fire scars form where heavy wood has rolled downhill and nestled against the bole of the tree. Even so, these stalwart giants have survived numerous fires over the centuries, stretching back to Medieval times. Slabs of tree trunks showcase their growth pattern and curved black swirls indicate the passage of fire, meticulously recorded in the giant's ancient woody rings, waiting for dendrochronolgists to catalog and count, while piecing together the fire history of the primeval forest.

Part of a Sequoia's mystery, and ability to survive these passing fires, lies in their egg-shaped cones which require intense heat to open. Once heated, oatmeal-sized seeds are released from beneath thickened bracts. As these seeds rain down from a height of several hundred feet, they land in bare mineral ash, the preferred substrate of young Sequoia seedlings. Like Lodgepoles, young Sequoias appreciate the sunny gaps created by the fire which punched holes through the canopy of neighboring trees, allowing sunlight to filter through.

Without fire, neither of these two species would survive. Ecological adaptations are a story that can be told over and over, in many different forest types across America. When we diligently work to stop all fires, humans alter the basic ecological underpinnings of many forest and grassland systems, some of which require periodic fires to survive in a healthy state. Though plants are called upon to adapt, humans rarely follow suit.

If people were as wise as fire-tolerant plants, they would actively contemplate the conditions needed for their homes and property to survive a wildfire event. Creating resilient landscapes is a necessity if we are to safely navigate the coming firestorm now headed our way like a freight train. That storm will not only be attacking landscapes in the U.S. for climate change is fueling intense fires in Russia, Australia, Israel, Canada and Indonesia too.

A recent tropical Indonesian fire released more carbon dioxide into the atmosphere from burning trees and soil than the U.S. economy spews out in a single year. In 2012 alone, 75 million acres of boreal forest went up in smoke in Russia as well. This begs the question: How can we learn to live more peacefully with fire? It isn't going away so how do we change the trajectory and our attitude?

This is where the human part of the fire story gets critical.

Creating Resilient Landscapes and Fire Adapted Communities

To lessen the coming storm, one of the first things we can do is reduce our carbon emissions and halt the overheating of our planet. Temperature is such a critical element on wildfires and the higher it zooms, the lower the relative humidity. Low RH levels suck moisture out of plants and dead wood, making them flammable. Fine fuels crisp first but if extended periods of dryness result, even large fuels such as logs will become kiln dry.

It was the long term drying out of large fuels in Yellowstone, in 1988, that helped to spur close to a million and a half acres going up in smoke. Small fuels registered a 2% moisture level with larger downed logs lagging not far behind. When large dead wood dries and burns, immense heat is generated and that can give a ground fire the surge it needs to climb into the treetops. Years of drought slowly cooked these logs till they were ready to burn.

Once lit, high pressure systems over the park held the heat and smoke of the fires close to the ground, encouraging them to flare up, even at night. Crown fires jumped from tree to tree, turning the skies black in the middle of the day, causing the sun to glow an eerie red. In one instance, six dozer-wide blades of removed fuel were jumped by the raging fire. The fire simply arced across the fuel free zone and ignited parched forests on the other side in a thunderous whoosh.

Aside from lessening our CO_2 output, humans also need to collaborate in order to return resiliency to our landscapes because fire knows no boundaries or political jurisdictions. In 2010, the Forest Service reached out to the Department of Interior and several other fire organizations and began formulating a cohesive fire strategy for the country. Prescribed fire was cited as part of the solution to the ever-growing fuel problem.

Prescribed fires eat fuel while killing off some trees, lessening the density of the forest. These fires are lit under precise conditions to help ensure the fire is containable. Weather parameters, favorable winds and fuel moisture levels are carefully selected and firelines are dug in advance. Trained staff are then prepositioned in key locations to keep the intentionally lit fires within acceptable territory. Prescribed burns are also one of the more cost-efficient means of restoring landscapes to health.

As the Rim Fire sped its way towards Yosemite National Park, Park Service fire crews began crafting a plan to slow the conflagration as it approached groves of mystical Sequoias. Although these giants are fire-adapted, they are used to frequent, moderate intensity ground fires not roaring infernos, spurred by decades of unburned fuel. With a ferocious firestorm headed their way, fire managers elected to light a back fire to reduce fuels prior to the arrival of the flaming front.

Saving giants, without burning down the rest of Yosemite, was the goal. Working to their favor were a series of burn scars left from previously conducted prescribed fires. Though these old burn scars had lightened the fuel load, managers set a series of sprinklers around the sequoia groves while carefully crafting their backfire. Simultaneously, they ran strips of fire through portions of the groves that were still fuel heavy to starve the coming fire of wood. In the end, they were largely successful. Fifty thousand acres of land burned in the park but damage to the ancient tree groves was minimized by using fire to fight the approaching blaze.

Clearly, prescribed fires and backburns can be a blessing.

Still, as is the case with all fires, there are risks and smoke. The more fire cycles that have been missed, the greater the fuel loadings, heat and intensity. Planned burns can escape and smoke is always a nuisance, but in fire-prone areas the smoke will come one way or the other. With prescribed fire, managers select ignition days to minimize smoke and can burn smaller chunks of land at a time to lessen the haze. It also gives firefighters the upper hand. Wildfires burn indiscriminately, often under the worst of conditions, arriving unexpectedly.

Nothing, however, is bullet-proof with fire. That means residents need to help prepare their own property by creating defensible space.

When wildfires blast through a WUI zone lined with homes, firefighters make critical decisions at rapid-fire pace regarding which homes they can protect from the coming flames versus those that are not worth the limited manpower. Often a single glance will tell a firefighter if making a stand on a home is worth the risk. Landing on the right side of that equation comes down to how much preparation the homeowner has taken on their own behalf. State agencies provide guidelines for removing brush, trees, small fine fuels and woodpiles away from homes and garages. Those guidelines should be followed and perhaps even enhanced if your property is thick with vegetation or on a steep slope.

Strategically locating propane tanks is another factor for not only does an exploding tank shoot metal shrapnel into the air, but it can cause a localized explosion of fire near your home. Selecting fire-resistant building materials is also a must. Though shake roofs are quaint, they flame up like matchsticks when embers fall on them. Draped pine needles on wooden decks are another hazard. Firefighters take enormous risks to protect homes, the least we can do is give them a fighting chance by preparing in advance.

In some fire-prone communities, agencies work together to create fuel breaks around communities…if there are dollars budgeted for such projects. For several towns near the Rim Fire, these fuel breaks proved invaluable. By lessening heavy fuel loads along miles of wooded land near Pine Mt. Lake, Groveland and Big Oak Flat, firefighters created enough defensible space to hold back the tide of fire. Starving the fire of fuel is the name of the game for it enables crews to quickly punch in dozer line while reducing the intensity of the blaze near homes. In some places, the breaks kept the fire on the ground rather than having it roar through the trees.

Fuel breaks should not replace individual efforts but they can be used to enhance the work done by homeowners. Helping fire-prone communities stay safe can involve prescribed burning, a rapid wildfire response and creating a resilient landscape by removing trees and brush along strategically located lines.

Preemptive measures can make WUI neighborhoods more fire-adapted. As firestorms sprint across the west, taking action in advance of the flames can make all of the difference.

From a budget perspective, it makes sense to shift agency dollars from extraordinarily expensive reactive measures to those tactics aimed at shifting the trajectory of fire, but right now some agencies are caught in a negative feedback loop. In 1995, 16 percent of the U.S. Forest Service appropriations went straight into fighting wildland fires. By 2014, that percentage rose to 42 percent. Since the agency's budget had remained flat, cash outflow was all going towards stomping out larger and larger fires, robbing programs aimed at ecological restoration, thinning and prescribed burning, according to a USDA report issued the year following the Rim Fire.

So just as humans must change their perspective on fire, so must agencies. Doing the same thing over and over and expecting a different result is what some might call insanity. Breaking the cycle of fire is going to require shifting dollars to proactive programs so landscapes can be healed, restored and made more resilient. Thinning overstocked forests and planting far fewer trees per acre following a fire are other tactics that could swing some public lands towards a healthier, more resilient state.

Thinking Outside of the Box

Thinking outside of the box can help agencies change the fuel-fire problem they currently face. Innovative people with a lot to lose monetarily are devising holistic plans that not only cut costs but also reduce the threat of fire. Meet Al Appleton.

Al used to run New York City's Department of Environmental Protection. He also oversaw the city's water and sewer system which was facing some of the same challenges fire often poses: threats to the water system. Removing ash and sedimentation from water systems is costly and just as Al was considering buying a new filtration system, he hatched a plan to invest $1.5 billion dollars into the health of the watershed that fed city pipes instead. The outcome of this brave new plan was that he saved billions of dollars in the long run, while investing in the health of surrounding forests. He was so successful, Boston, Seattle and Portland followed suit, with all of them saving millions of dollars in avoided water treatment costs as a result.

In California, 50 percent of the surface water flows from National Forest lands. That makes these forests major producers of one of the most valuable resources around…water. Taking care of these watersheds, and the meadows that dot their surface, should therefore be a primary mission, for they both play a role in the hydrologic functioning of a forest.

Think of it this way, watersheds are the forested corridors along which rivers traverse and meadows are the catchment basins that capture the flow. Meadows function like a sponge, soaking up vast amounts of water, slowing it down, and then trickling it downstream slowly, over time. Water from spring snow melts is effectively portioned out into hotter, drier months of the year by this very action.

Healthy meadows also filter flood waters, straining out ash and other sediments. Following a fire, meadows act like a sieve by trapping soot and other pollutants, insuring that the downstream flow is relatively free of debris. By slowing water down, it also reduces the chances of post-fire flooding which is always a threat when large areas of soil have been badly burned. Well-functioning watersheds and meadows could save millions of dollars downstream, for this very reason. That's why putting money into restoration makes so much sense. The problem, however, is huge.

Out-pacing degrading landscapes in California alone, would require restoring six to nine million acres of forested land over the next two decades. That work would need to entail reducing the density of trees in waterways, possibly through using prescribed fire, while restoring meadows to their natural state. Healing the land would also help thwart the trajectory of catastrophic fire. Since critically burned watersheds take roughly 7 to 10 years to recover, while plants slowly recolonize barren soils, restoring forested waterways would effectively lessen downstream sedimentation problems that are clogging reservoirs in many states, following huge fire events.

If we desire a future riddled with fewer catastrophic fires, we need to heal our landscapes, remembering that the value of these lands is not so much in their timber, nor in the water they provide, but in their health, vibrancy and resilience. For the 44 million people who have homes tucked in the WUI, a vested interest in healing these landscapes should be paramount. Water districts that rely on stream flows coming from these forested mountains should also be concerned. Good land stewardship needs to become an everyday project that we all work towards, for everyone stands to benefit.

The Coming Flood

If we elect to ignore the true value of watershed resources, we will pay the price in coming floods.

This is true because beneath all of those forested waterways lies one of the most forgotten resources of all: Soil. Soil is a complex ecosystem squirming with nutrients, billions of bacteria, protozoans, insects and nematodes. Unbeknownst to us, our earthy matter houses a quarter of the biodiversity of the planet within its layered realms.

Think of soil as the living skin of the earth. Like meadows, soil helps to filter water while banking seeds that spring into plants following fire. As these plants sink roots, they help restabilize burned hillsides, but that often takes many years. When unburned, soil is covered with leaves, twigs and plants that bind our world together, keeping our feet on solid ground.

So what happens when a fire whisks through an area, burning even the soil? Intensely burned soil can slide, slump or even dump many things, causing massive debris flows to cut loose on the landscape. This can impact entire hillsides or start in small channels, but as more rain falls, the end result is that large gullies cut into the landscape, funneling rain down storm shoots, much like gutters funnel water off your roof.

This happens because hot fires vaporize fatty acids in the organic matter of the soil. These superheated compounds then sink into the pores of the soil causing clogs. When that fatty layer hardens, it creates a slip and slide like surface to the soil itself. That slippery surface repels water, making it hydrophobic or "water-fearing." Since the water can't sink in, it slides off, causing major water runoffs. As water glides over this surface it tends to drag ash, soot, sedimentation and burned debris in its wake.

On steep mountain slopes, if rainwater falls fast enough, it can create debris flows, flash floods and mudslides.

Huge rocks and even trees can get swept up in the sea of blackened mud flowing off these burnt mountains, cutting deep stream channels as they careen downhill. Tides of dark, sooty water can rise rapidly creating floodwaters that take out houses, toss cars about or inundate narrow river canyons. These are also the same floods that dump literally tons of sedimentation into water reservoirs downstream.

Because these post-fire events are so damaging, scientists are often called upon to create soil burn severity maps immediately following the wildfire, to help identify trouble spots quickly so they can mitigate damage to those living downstream. Satellite imagery and infrared are used to create these maps but ground-truthing is an essential part of the process. Poking around in the dirt to determine how much organic matter has been lost is critical for it tells scientists how deeply the fire's heat penetrated into the soil horizons below. The more clogged the air passageways of the soil, the greater the damage and chance for catastrophic runoff.

Soil studies are conducted to help mitigate risk to property owners before trouble arrives by stabilizing the hardest hit areas with straw, installing culverts and water bars and by taking other proactive measures. Post-fire rain events can pass over an area quickly, but the damage they cause can cost lives and tremendous amounts of money to repair. They can even whisk away the seed bank, making it harder for plants to recolonize the site.

Losing a great deal of soil over a broad area is always a huge concern because it can snatch away the nitrogen housed in the upper foot of the soil. This is vital because nitrogen fuels plant productivity and without out it, plants have a hard time reestablishing. That's why plants such as lupine and deer brush often inundate burned sites. These plants have the ability to fixate their own nitrogen using specialized root nodules, allowing them to colonize nitrogen-deficient areas. That gives them a leg up over plants that can't grab nitrogen from the air and essentially relocate into the soil beneath their roots.

Though emergency rehabilitation teams are efficient at stopping the worst post-fire problems from occurring, when huge swaths of land are impacted, it is impossible and too costly to treat the entire landscape. That's why soil becomes such a critical matter when it burns. It is also why downstream reservoirs are so often inundated with ash flows.

When the earth is stable beneath our feet, we largely ignore it but when it cuts loose and starts moving, it grabs our riveted attention. Rather than paying to clean up the mess, maybe it's time to heal our landscapes instead and plant our feet on more solid ground.

Doing so will necessitate that we accept the natural role that fire plays in many ecosystems. Likely, it will also require using a variety of methods to lessen fuel loadings to protect communities. In many wilderness settings, it might be wisest and safest to simply let fires burn on more public lands, while herding it away from populated areas, so it can eat up fuel.

On forests that grow trees as a product, it is essential to thin overly dense stands, making them more resilient, but doing so requires that budgetary dollars are set aside for such activities and not used to squelch the next emergency fire. Following a fire, foresters need to calculate in the effects of our changing climate. Sites that are warmer and drier than in the past should be planted with far fewer trees, taking into consideration tree types geared towards hotter temperatures and water scarcity. Likewise, they could also consider letting the forest simply reseed itself. Pines have been reseeding burned mountains since time began and they naturally thin their numbers according to the holding capacity for that site.

Certainly, solving the problem will demand re-prioritizing budgets and projects. Shifting ourselves from a reactive state to one of proactivity is going to take effort, time, money and manpower. Reexamining policies that box us in will be a prerequisite.

Or, in the absence of change, we can simply wait the for coming storm.

SEXY CIGARETTES, NUCLEAR FALLOUT AND FAKE NEWS

Big Tobacco

What does a flapper seductively smoking a cigarette, billboards on nuclear testing and tweets about fake news have in common with misinformation on climate science? All of them are brazen examples of propaganda campaigns aimed at propagating, or growing, deceptive information via the use of news stories, sign posts, movies, social media sites and advertisements.

Persuasively shifting the public's opinion regarding an institution, person or cause is the desired outcome of these misleading campaigns. Another way to think of propaganda is a means of leading the sheep astray. If done convincingly enough, these repetitive, credible-looking ad campaigns have the power to sway the masses in the wrong direction.

Consider the sleek ads that targeted women in the late 1920's, convincing them that smoking was sexy and alluring. Smoking was seen as a status symbol, evoking images of modern women striving for emancipation and freedom. The "Torches of Freedom" cigarette ads broadcast in the U.S. were deemed vital to the American Tobacco Company for the president reckoned that recruiting women to the ranks of the smoking would "be like opening a gold mine" for the company.

Women were ready for change in America. As housewives entered the labor force during World War I, they gained a sense of independence and rebelliousness. Smoking stoked the image of women breaking boundaries and they lit up in record numbers, driving cigarette sales through the roof. Phillip Morris even ran a lecture series to inspire women to strike the right pose while sucking down cancerous smoke.

The "Torches of Freedom" ad campaign was so successful here that the tobacco companies began pushing the same agenda across the seas. Each country embraced the idea in slightly different ways, according to their culture. In Japan, cigarettes made women feel unique, allowing them to stand out in a culture steeped in tradition. Smoking for Indian women became a status symbol. By dressing smoking models in western garb, puffing became a glamorous habit linked with upward mobility.

SACRED GROUND | 133

Just as these ad campaigns had worked at home, they were also widely successful abroad. Worldwide, more women were puffing, driving up tobacco sales, making company CEOs richer at the expense of womens' health.

When people began questioning health-risks associated with smoking, the tobacco companies went back to the drawing board. Throat-scratch ads appeared along with posters that showcased doctors firing up, clearly showing there should be no concerns with using tobacco. If doctors were smoking, naturally it had to be safe…at least that was the implied message. (http://content.time.com/time/photogallery/0,29307,1848212_1777642,00.html)

Shameless commercials showing pregnant women firing up were some of the more salacious ads promoted. (https://www.pinterest.com/pin/39336196715611469/) With such a smooth taste, what expectant mother wouldn't want one?

Despite volumes of information pointing towards severe health risks associated with smoking, tobacco companies continued to pour out propaganda. Why? Propaganda works; the bottom line was proving that. Say something long enough, with enough conviction, colorful pictures, the right images and fake information to back it up and people can be bullied, coerced, misled or even gently guided down the wrong path. Consider these ads an early version of a tweet storm.

Tobacco companies packed cigarettes with enough nicotine to make them hyper-addictive, even changing the chemistry of the brain, so smokers craved their product with an endless, unquenchable hunger. Corporations can be very evil-minded when it comes to making vast amounts of money, working against the better interests of society while fueling their avarice, padding paychecks in the process.

Nuclear Fallout and National Security

Though money drove the tobacco industry, promoting a cause was the aim of the nuclear testing campaign that flooded western states with nuclear fallout for roughly 40 years. Building and then perfecting nuclear bombs was an American priority following Japan's invasion of Pearl Harbor. By then, the cold war with Russia was underway and there was concern of a nuclear attack. Civil defense films and posters tapped into our deepest psychological fears, warning us of imminent battles. (Google: historic examples of U.S. propaganda signs for nuclear testing)

Billboards popped up warning people to keep their silence if they valued national security. Meanwhile, bomb after bomb was tested in the deserts of Mercury, Nevada between 1951 and 1992, propelling plumes of radioactive dust and fallout over the southwest. Bombs larger than the one dropped on Hiroshima, including "Dirty Harry" were detonated at this site, which became known as one of the most heavily used testing grounds in the world. Radioactive ash was so thick in St. George, Utah it became known as "Fallout City," but people were encouraged to keep their mouths closed as a matter of state defense. Speaking up was considered "unpatriotic."

People living downwind of the site began to pay the price. For though films such as Bert the Turtle (https://www.youtube.com/watch?v=IKqXu-5jw60) encouraged students to duck and take cover under their desks in the event Russia nuked us, some local school children were actually told they could go outdoors to watch the flash and ballooning mushroom cloud so long as they wore sunglasses. In either event, the advice was ill-conceived for neither set of protocol was adequate for keeping Americans safe in the event of nuclear warfare, or testing for that matter.

Children, adults and livestock exposed to the radioactive waves and dust began to fall ill in unnaturally high numbers. Cancer rates escalated along the fallout path and concerned women tallied deaths in their neighborhoods, keeping meticulous records of the illnesses along with the people impacted. Downwinder stories poignantly document what occurred in neighborhoods in Utah as the radioactive dust settled. View: https://www.youtube.com/watch?v=N2XXlO240ac and https://www.youtube.com/watch?v=y-GeccG7zOs.

Like the tobacco companies, the Atomic Energy Commission intentionally withheld information from the public, to promote their own agenda: National Security. In other instances, our government's knowledge of nuclear energy was quite lacking but they provided what they considered to be reasonable safety protocol. Campaign ads aimed at getting the public onboard the national security agenda were dispersed widely. (https://www.youtube.com/watch?v=yE-nji8-1Ko)

Meanwhile Geiger counters jumped in Southern Utah when deployed in schools to measure radioactivity levels emitted by children. Additionally, approximately 5,000 sheep died in the spring of 1953. These deaths foreshadowed the fate soon to befall the sheepherders, for they too were caught off-guard, working outside when the bombs exploded mid-air. Many families living downwind of the test site did not fare well, suffering one type of cancer after the other during the course of their lives.

Fallout from the bombs started showing up in milk supplies across the west, for the nuclear fallout reached far and wide. Despite the radioactive milk, some schools were required to serve milk at school lunches. There were even extensive ad campaigns aimed at boosting milk consumption.

Eventually, the tireless efforts of a handful of women pushed the downwinder story into the forefront. As time passed, their meticulous epidemiology studies became hard to refute. Personal stories of suffering started hitting the main-stream press and pressure was exerted on the government to stop nuclear testing at the Mercury, NV test site, at least temporarily.

With a discerning, involved population, truth has a chance to win out over fake news. Engaged citizens, willing to point out fallacies they see is critical to maintaining veracity, healthful conditions and a working democracy that is responsible to the people.

Speaking power to truth is necessary in our country. When women raised their voices against the deaths that were occurring around them, a measure of justice was won for some downwinders via settlements and healthcare assistance.

The Era of Fake News

Legitimate news stations, in recent years, have suffered numerous, pernicious attacks with many of the barbs coming straight from Washington D.C. Never before has such a relentless attack on the free press ever been seen in the history of our country that I have witnessed.

The reason this "Fake News" ad campaign is so dangerous is that it is repeated consistently on TV as well as twitter. Scripted messages have even been read on stations all across the country in an attempt to brainwash Americans, with journalists parroting messages written by politicians, employing no judicious thought of their own, while airing strictly controlled propaganda. When I saw this happen on our own networks, I thought perhaps we were watching an English version of Russia Today rather than a major U.S. news station, where freedom of the press and speech is supposed to be a cornerstone of our country.

Worldwide, this is how dictators take control of the masses. First, they discredit legitimate news sources, insulting individual journalists, while cannily crafting their own message, which ironically turns out to be based on anything but the truth. By labeling real news as fake, politicians plant seeds of doubt in the minds of many.

Deceitful as this practice is, it sets up the naysayer for delivering what they wanted to promote all along, their twisted, self-centric version of the "Real Truth." Unfortunately, this version of the truth is usually self-serving and not always grounded in facts.

Promoting his version of "real news" was the reason Hitler started the Ministry of Public Enlightenment and Propaganda. White-washed versions of the Nazi Party were delivered to the world via the ministry as it took full control of all information disseminated by the party, including visual arts, movies, theatre, radio announcements and even the news. People were warned of a Communist uprising and their worst fears were tapped by politicians who then stripped them of their civil liberties under the guise of protecting them from pending threats.

Printing presses were seized and a concerted effort to drive out competing papers was undertaken by the Nazis. As market prices fell to cutthroat rates, the Nazis scooped up a vast majority of the country's newspapers. Racially pure, yet inexperienced journalists were then hired by the Third Reich who indoctrinated their writers and editors with a stern hand, ensuring that only information approved by the party was disseminated to the masses. Veering from the approved path could cost a journalist a job or even a long, horrible stint in a concentration camp. Experienced journalists who were not loyal to the party fled.

Under Mussolini, censorship and propaganda were used to shape the minds of the Italian people. Though Mussolini, a former journalist, swore he would protect the freedom of the press, he also refused to print anything that was contrary to national interest or that went against his version of the "Fatherland."

Journalists were used to indoctrinate the masses. Investigative journalists, who applied scrutiny, fact-finding, or who wanted to voice alternative ideas were summarily squashed. Militant journalists, willing to follow party lines, were the only writers Mussolini had a use for on his state run papers. To bolster his fascist position, he proposed comprehensive legislation to govern censorship. Bookstores that sold contraband ideas were raided. In time, Mussolini collaborated with the Third Reich to garner more ways to lock down the free flow of information.

Climate Science

Controlling information is one of the more malignant and insidious ways that people, corporations and governments employ to keep the public in the proverbial dark. This can take the form of removing vital information from websites, yanking funding, placing gag orders on agency experts, displacing scientists from their jobs into meaningless positions, shutting down organizational communication and outreach strategies while providing an alternative to the truth, something most of us call misinformation.

Climate science has been subjected to a host of these illicit communication strategies in order to obfuscate and muddle the truth. Since we now recognize propaganda campaigns for what they are, the question becomes, why? Why would anyone want to keep Americans in the dark about climate change, something that is so important to our future, the health of our National Parks and the plight of humanity?

To answer these questions, you have to follow the money…not only corporate money, but also the stock portfolios of politicians along with campaign funds, payoffs, kick-backs, post-government positions and other perks laid out on the government "all you can pork down" buffet table. For no one would go to such a great extent to malign the science unless it was extremely profitable, while also advancing a political agenda you were already vested in.

Better yet would be a propaganda campaign that had it all. By tapping into the fear of the people, you could push a national agenda that would personally make you grossly rich. These days, that is a politician's hottest and most lucrative dream for the world gobbles up roughly 42,000 gallons of oil every second the planet spins on its axis. All you have to do to money-grab your share is to push the right agenda on The Hill.

The only problem is that the research that credentialed climate scientists are collecting overwhelmingly points to the fact that C02 emissions are causing enormous environmental problems worldwide. Floods, droughts, atmospheric rivers of rain, sea-rise and rain bombs could unleash civil unrest on an unfathomable scale if we don't turn things around soon.

So who could possibly want to profit from such a short-sighted and destructive set of circumstances? Big Oil,

Big Gas, fracking companies, major stockholders and unscrupulous politicians.

Let's start with the corporations. Somewhere in our brief history as a democracy, America has ceased to be "We the People" and turned into "Thou the Corporations" instead, leaving the average person struggling by while CEOs profit immensely. Dark money from these companies pour into our campaign system, some of it possibly being white-washed in foreign countries before being pumped back into their party of choice, effectively buying politicians who are then expected to toe the company line even if that means promoting agendas that are harmful to the vast majority of the people.

Backing these corporations are special interest groups who spur the agenda of burning more fossil fuel. Huge ad campaigns, climate science denying books and even stilted natural history museum displays funded by oil companies are being used to sway public opinion in the wrong direction regarding climate science.

One study by Peter Jacques discovered that between 72 and 87% of climate change denying books can be linked to think tanks funded by those who will profit richly from the misinformation being spread.

If you are interested in following the money trail, read Naomi Klein's book, "This Changes Everything: Capitalism vs. the Climate." Knowledge will help you to unfund the organizations, think tanks and parties who are hurting our country for profit. Like the tobacco industry, these companies know full well they are causing severe harm but if they can obscure the facts long enough, they stand to ring out billions more in profit while derailing attempts to switch to more sustainable energy while we still have the resources to effectively make the change.

Anyone trying to get ahead these days buys stocks for the average person has no hope of retiring without investing. With the severe environmental concerns facing us today, we all need to make tough decisions regarding which companies we truly want to back. The 350.org website contains a wealth of knowledge regarding how you can go green while divesting yourself of the companies you no longer wish to support. Making the switch is up to us…individual stock holders can undermine harmful companies by yanking their funding.

If climate change concerns you, pressure needs to be exerted on your elected officials to ensure that your voice is heard. Al Gore's book, "An Inconvenient Sequel: Truth to Power" paves the path for getting politicians to hear your concerns. You can start by finding out who your representatives are at usa.gov/elected-officials. Track how they are voting on key issues at scorecard. LCV.org and call them often at 202-225-3121 if you aren't happy with how their scorecard is stacking up.

Though you could question why and how Al Gore lost his run for the presidency, I would argue that he had a higher mission and that was starting up his climatereality.org project. Since 2006, Mr. Gore has been taking science to the people, helping the average person to understand the implications of increasing CO_2 emissions in our atmosphere.

In Conclusion

My greatest concern is to save our National Parks while also promoting an environment that can peacefully sustain humanity. In doing so, I have tried not to hammer any particular political party because in truth, we all need each other in this country if we are to create effective change.

Propaganda, half-truths and outright misinformation have to be recognized for what they truly are. People deserve to have access to all types of information so they can make informed decisions on their own regarding matters of health, happiness and public safety, including information on climate change and climate science.

Despite what we'd like to think, government officials often do not have vital information needed to make sound, educated decisions when operating within an isolated bubble. When these same people refuse to listen to (or fire, reassign or unfund) highly trained and educated people that can clarify matters for them, red warning lights should flash for those of us watching. Educated professionals are often the first to be attacked in authoritarian countries. Hiding the truth is easier when you bury the science along with the people plunking the puzzle pieces together.

Miscalculating a position on climate change could prove fatal to humanity. With so much hanging in the balance, we need to engage the brightest of minds on our behalf. When we vilify climate scientists, forcing them out of key jobs, in order to hide the truth from the people, we have started down a dangerous path indeed. When those actions are followed up by attacking the press and removing scientific data from websites, we have slid even further towards a very scary political state that looks more like Nazi Germany than the United States of America.

Changing the tide of a miscalculated government position requires decisive and relentless action on the parts of people who care, for only when a critical mass is reached will politicians start doing the right thing. Unfortunately, too many of them are owned outright by special interest groups who are stuffing their pockets with greenbacks.

As climate changes unfold around the world, we need the free press to keep us informed in an unbiased manner, but choose your news sources discerningly.

Educating ourselves on matters of concern is enormously important so we are not led about like docile, nucleated sheep. Back this up by reading a variety of books and magazine articles, talking to experts in the field and watching films from trusted sources. This is vitally important because we have entered a new age, wherein we have to sift through what is true and what is false extremely carefully.

Fake news abounds, but it may be coming from sources you least expect. Ask yourself, who stands to profit from keeping people in the dark? What is their ultimate agenda?

Keep your eyes open America, we have so very much to lose. Ninety-seven percent of climate scientists are in agreement: The planet is warming and they believe that human activities are almost certainly the cause. Can we afford to make decisions in the dark?

With so much agreement amongst scientists, it makes me wonder what books are our politicians reading? More importantly, why are they failing to act?

HOPE IS ACTION

Hope exists where there is action. When caring people come together to make a difference, anything is possible. All around the world, people of all colors, nationalities, religions and backgrounds are joining hands to face our Rock and the Hard Place: Battling Climate Change. There is no more worthy a cause to align yourself with and as we pitch in and see change, we will realize that action also creates hope.

Here is a list of things you can do to make a difference now. Bear in mind that some of these tasks will make sense for some of you, while other action items will work for different segments of the population. I suggest a variety of solutions because we all need to respond in our own fashion. We are a melting pot of great and varied people so rather than focus on something that doesn't work for your mindset, realize it may work extremely well for someone else. The goal is to dig in where you can. As you do, consider the enormous difference you are making for the world.

Action List: Things You Can Do Now to Help

- Read Fred Pearce's book "With Speed and Violence." He is an experienced journalist who is well versed in climate change issues. Not only is his book a page-turner, it is also an excellent primer for understanding the scope of what is occurring around the world.

- Join 350.org. This grassroots organization mobilizes people all over the planet to combat climate change. They also encourage other organizations from divesting themselves of the fossil fuel industry ([https:// globaldivestmentmobilisation.org](https://globaldivestmentmobilisation.org)). Several religious organizations are already backing this movement on behalf of all people and I encourage your church group to join as well.

- Our elected officials have forgotten a key point: They work for us. Too many Americans have become complacent about how our government is or is not working. The less we pay attention and act, the further off point our democracy becomes. Like climate change, it has become a runaway freight train. It is time to organize, engage and put the train back on the right track. If you care about issues such as health care, clean water, climate change, protecting our environment, decent wages and affordable tuition for our children, it is time to remind people in Washington who is in charge. **Call this number every time you feel you are not being represented well: 202-225-3121. Share the number with like-minded friends.**

- Put a solar water heater in your house. You could save up to 30% of your home's carbon footprint by doing just this one thing.

- Join Al Gore's Climate Reality Project at https://www.climaterealityproject.org and start sharing information with your community. Al's leadership and training class will teach you what you need to know to inform others: https://www.climaterealityproject.org/training

- Dump stocks that fund companies you no longer feel good about while investing in those that bring hope for humanity, such as renewable energy sources.

- Read "The Weather Makers" by Tim Flannery. He is a big science guy and his book goes into a lot more depth for those of you questing for more information on climate science. At the very end of his book, he also lists action items you can work on to help create positive change.

- If you have the means, consider putting solar panels on your house and buying an electric car that you can recharge using solar panels. In every way you can, reduce your carbon footprint. A good starting place is to figure out how much carbon you are now using, by plugging information into this calculator: https://www.carbonfootprint.com/calculator.aspx

- Close family and friends can prove to be some of your most difficult encounters when it comes to talking about climate change, especially if they fall well into the denier camp. Rather than shun these encounters, use the information posted on John Cook's Skeptical Science website to intelligently swing their opinion in the right direction: https://skepticalscience.com/posts.php?u=1

- Put pressure on every administration to ensure that we are funding programs that will help us to see trouble coming. If scientists hadn't found the ozone hole that opened up over Antarctica a few decades ago, we might not be here now. Rather than embrace the good things science can do for us, the Trump administration has chosen to bury its head in the sand while defunding critical programs aimed at monitoring our atmosphere and ocean waters. We cannot afford backwards, climate-denying politicians…too much is riding on this issue such as the health and well being of your family.

- Fund the National Park Foundation, which supports our National Parks. Talk to your elected representatives in Congress and tell them about the impacts National Parks are already facing from our changing climate. Climate change stories abound in our parks and the problem is widespread and pervasive, affecting many species. The earth is speaking to us and we need to start listening. If your elected official isn't working to solve the problem, you need to get them engaged.

- Strangely, our countrymen and women seem to be at odds with our federal government when it comes to climate change. That makes me wonder who is making money in Washington to push fossil fuels at the expense of us all? Since the swamp is swampier than ever, states and local communities need to take action to undermine policies aimed at harming us. Example: On a state level, New York and Vermont passed laws to make fracking illegal. After seeing what happened to Pennsylvania, it became obvious that to protect their watershed, they needed to keep the oil and gas industry at bay.

- Consumption levels must be lowered and Americans need to lead the way. We have too much stuff and producing it takes gas and oil at every single turn. Consider shopping more at Goodwill stores, Deseret Industries and other organizations focused on repurposing items already created. If you doubt that we own too much (and I am no exception), take a look at the book: Material World: A Global Family Portrait, by Peter Menzel. It is a shocking, eye-popping pictorial review of just how much we own compared to other people around the world. Share.

- Freedom of the press must be protected. We need investigative journalists who are free to research and report the truth on a variety of issues, including climate change. Anytime you hear that a swampy Whitehouse is attacking journalism, or is placing gag orders on climate change scientists while removing stories about climate science from the internet, you need to complain loudly. Keeping people in the dark about matters of great importance is not only wrong, but in the case of climate change, it is also extremely dangerous.

- If it is not against your religion, strongly consider not having children. Over-population and over-consumption, are two of the key issues driving climate change. Voluntarily, we can reduce population or we can wait for climate change problems to do it for us in the guise of insect borne diseases, famine, ecosystem collapses, resource wars, watershed problems and from extreme weather events.

- Be energy efficient at home. Turn off lights and do your best to control your thermostat during all seasons. Unplug electronics that suck juice for no reason and purchase appliances known to be efficient, such as Energy Star.

- Switch to a vegetarian diet. By avoiding meat and dairy, not only will you feel much healthier, but you will help to stop about 18% of our greenhouse gas emissions. Growing a garden and eating food locally produced also helps.

- Pressure needs to be exerted on our politicians to hire ethical leaders for the Environmental Protection Agency. When we deregulate the EPA, we are setting ourselves up for health and climate problems down the road. Since we are at a crossroad, we can no longer afford to place oil and gas industry people in this position if we want them to work on behalf of the people.

- Stay informed by reading such sites as:

  https://climate.nasa.gov/newsletter_signup/
  http://www.eesi.org/newsletters/climate-change-news-ccn
  https://insideclimatenews.org/newsletter/welcome

http://climatenetwork.org/about/members
http://climate.calcommons.org/article/staying-current-climate-change-newsletters-listservs-and-updates

- Beware of "climate change" sites funded by think tanks that take dark money from the oil and gas industry. Naomi Klein's book "This Changes Everything: Capitalism vs. the Climate" points out a number of organizations and institutions that are purposefully hiding, and even lying about the truth regarding climate change. Much like Big Tobacco, who did everything within their power to keep people puffing cigarettes despite the health hazards, these organizations are more than willing to hide the truth to line their own pockets, while causing extreme harm to humanity and the planet.

- Join climate marches.

- Fly on airplanes less.

- Engage the youth of America on the topic of climate change. Many of them are unaware of the problem but it is our youngsters, and their families, who are going to feel the full brunt of the issue, in the near future. As we have seen from the Parkland students, they are an energetic bunch and they are connected so tap into that energy to spread the word.

- Use social media sites to get connected and spread the word in a positive way. Not everyone reads books these days but data bits, coupled with compelling photos and stories, can help to catalyze the youth of America.

- Pope Francis has taken up the climate change cause on behalf of all people. Speak to the hearts and minds of our other religious leaders to do the same. Climate change is very much a human issue. If you consider the ramifications it could cause around the globe, it is also a moral problem we need to solve. Religious leaders can help lead the way.

- Consider funding projects promoted by Leonard DiCaprio's Foundation. He has been a spokesperson for climate change for many years, using his position, money and fame to do good things.

- Thank people like Michael Bloomberg and other philanthropists who generously give to make good things happen on behalf of us all.

- Recycle. Insulate. Drive less, walk more.

- Plastic bags take a lot of oil to produce. Quit using them and/or work to pass laws that make them illegal.

- Friends of the Earth (https://foe.org) launches many vital petitions related to environmental concerns such as offshore drilling and climate change. They also help to keep political stories in the news so people can see what politicians are up to.

- Sign petitions often for it is one way that politicians get a sense of the things that matter to the American public. The more of us that sign, the better.

- Consider what is at stake in the environmental game of Jumanji we are currently playing with the climate and encourage all people to take up this cause. Our future depends upon it.

- Build a tiny house. Not only does this minimize the resources needed to build your home but it will keep you from stockpiling commercial goods, all of which take enormous amounts of gas and oil to produce. Meanwhile, you can bank your house payment, placing it into a retirement or vacation fund.

- Encourage students to learn science in school. Currently, we are math and science deficient in this country and perpetrating myths is a lot easier when people have a hard time grasping issues. By starting early, we give the people of this country a head start in understanding matters that are going to affect them greatly.

- Support programs that can help workers engaged in the extractive fuel industry to retrain because I do think it is very important that we don't have people lose out in the transition. Not doing something you've always done can be extremely daunting, unless you are given a healthier, more stable option. It would be great if someone could help those displaced by the transition by setting up a fund to retrain them in solar or wind-powered technology.

- The opposition is getting VERY organized and is placing enormous pressure on lobbyists to sway things in their direction. If we don't do the same while maximizing our potential as a diverse group of concerned citizens, our voices will not be heard.

- Campaign laws need to be changed. When corporations are given such a strong voice, the concerns of the people will not likely be heard. If you have political law savvy, this would be a great place for you to engage the problem at its core. The health of our citizenry should take priority over the wealth of corporations who are not working in our best interest.

THE ROCK AND THE HARD PLACE

Writing about fracking, ocean acidification and deadly hurricanes crashing our shores is unnerving because it forces me to consider all that is at stake on our planet right now. Admittedly, climate change makes me fearful on many levels. Emotionally, it is devastating because I realize that places I love could be unrecognizable to my own eyes in years to come. From a human perspective, it is horrifying as it has the potential to unleash enormous suffering around the globe. Politically, it is a nightmare because it requires value shifts from industries and the politicians who flagrantly support them. Our health and well-being need to take precedence over profit margins, which many companies seem to value above all else.

Although climate change is an insidious, overwhelming and seemly unconquerable problem, if you approach it from a spiritual perspective it then becomes the greatest opportunity ever presented to humanity. Consider it a divine test...one that requires us all to join together, to set our differences aside, to work across the normal lines that typically divide us such as race, socioeconomic levels, nationality, party affiliation, gender and religion. It will require us to dramatically alter our value systems, placing saving the planet and humanity above individual agendas and greed.

Despite great odds, it presents a test I believe we can pass. It all comes down to hope and a fierce determination to succeed. Deciding that failure is not an option is one of the first attitudes we must embrace. Then we need to join hands, get busy, pray and take decisive action **NOW**. Halting climate change is possible and I don't mean this rhetorically. In my DNA, I believe it to be true, for I know a few things about being in an unwinnable, untenable, impossible and hopeless position. Fourteen years later, I have been delivered to the other side, cancer-free. My healing journey made me a far stronger person than I ever would have become without the challenge.

You can tackle your worst enemy head on and survive to tell the tale, it is entirely possible. Along the way, however, you will have to confront your darkest fears and proceed anyways, despite the gloominess and unknowns. You'll be forced to decide what's important and what you have to release like a bag of rocks weighting you down. Your faith will be tested like never before. In the end, I will say this: God helps those who help themselves.

I wasn't healed from cancer miraculously nor overnight, I scratched and clawed and cleansed and ate, sweated, prayed, swallowed ghastly stuff, meditated and fought my way to health one day at a time. Absolute determination, coupled with hope, faith and a relentless desire to live made me successful. I met my own personal rock and the hard place and I decided to fight. That battle made all of the difference in my life.

My own experience, and that of other survivors, is what gives me hope for humanity and their ability to solve the challenges posed by our changing climate. It all comes down to the immensity of the human spirit and how we choose to confront the problem psychologically. Determined fighters win. People, who decide against all odds to overcome, will find a way.

Those who perish are the ones that give up at the onset of trouble. Depression, hopelessness, lack of action and passively waiting for help to arrive seal self-prescribed fates for many. We are more in charge of our own fortunes than we realize. Lament the looming changes thrust upon us by our changing climate and postpone action…or fight like hell? I subscribe to the latter method and am hoping you will too.

During my illness, endless books and movies about survivors fueled my spirit. In consuming them, it became evident that certain traits, actions and psychologies predispose people to overcome great odds. When facing issues presented by climate change, survivors can teach us so very much.

Consider the book "Alive, The Story of the Andes Survivors," wherein a plane loaded with soccer players crashed into the Andean Mountains in the dead of winter. The crash killed many of the passengers outright but the others stranded on that stark ice-clad mountain were faced with a number of harsh choices, the greatest of which was whether or not they should eat body parts from those who had already passed, some of whom were friends.

Extreme deprivation, frigid cold, isolation and an avalanche that struck the downed aircraft forced their hands. After a great deal of prayer and contemplation, the decision was made. The dead would be used to help the living. Despite the sustenance provided by cannibalism, survivors suffered harshly and continued to waste away over the course of the winter.

Hopelessness set in as one survivor after the other died, dwindling down the numbers of men sitting on that lonely mountain, waiting for a rescue party to find them, saving them from their fate. Finally, two men realized the rock and the hard place had arrived. Help was not coming. They had to hike off that steep mountain in horrible physical condition or resign themselves to dying in the guts of the broken plane.

By harnessing a ripped off airplane door as a sled, Nando Parrado and Roberto Canessa slid down the vertical mountain, beginning a ten day journey that ended in rescue efforts for the remaining crew. Against all odds, they overcame their severe struggles.

In the movie, "Touching the Void," two mountain climbers were bagging peaks in South America when one slipped and fell into a deep crevasse. His friend, Simon, tried desperately to pull Joe out but after a while it became obvious that if he did not let go of the rope, they both would die. Simon thus did what he had to do: He cut the rope.

This sent Joe cascading further down into the bowels of the ice crevasse. Along the way, he broke multiple bones before coming to rest on a ledge, wedged between walls of ice. Joe had met his rock and the hard place. Do nothing and he would freeze to death in an isolated crevasse, in the interior of a jaggedy mountain with nothing to witness his death but walls of solid ice.

Climbing back up the crevasse was not an option but down below the crack was what looked like a sliver of light. If he slid off his ledge, letting go of the only solid thing he had in his life, he opened up one very important thing: possibility. There was no telling what was at the bottom of the tunnel he was stuck in, so, he did the only thing he could…he slid. At the bottom of the tunnel, Joe found a way out of the crevasse.

Finding a way out of the crevasse, however, was not an end to his tragedy; rather, it was the beginning. With broken leg bones and multiple other injuries, Joe had to cross a huge expanse of ice-covered wilderness by dragging himself on his knuckles. The grueling journey took days. Miraculously, he arrived back at base camp just as Simon was packing up, readying himself to leave the area behind, believing his friend to be dead, frozen in an ice crack.

"Endurance" is a harrowing tale of survival on the seas and frozen landscapes of the Antarctic. Ernest Schackelton left for the South Pole in 1914 on an exploratory trip. Just shy of his destination, drifting ice hemmed the ship in, leaving his crew stranded for months on end. With supplies dwindling, severe rations were put in place but an attitude of hope kept the men alive.

Straining sounds moaned into the ice-cold night as Schackelton's ship got squeezed by ice floes, eventually necessitating that the men abandon the pressurized ship, taking up the open ice as their home. After months of eating seals and surviving horrific cold, a few men boarded a small boat and crossed wind-tossed seas, seeking help on the island of South Georgia. The crew had finally met their rock and the hard place. To get beyond it, they had to risk a nearly impossible journey across rough and frigid waters.

Excellent leadership, teamwork and a sense of hope kept the men's spirits aloft during their ordeal. One trial after the other was faced and overcome under conditions that would have crippled less strong souls.

In "127 Hours" a young man literally met his rock and the hard place in a far-flung gorge where he went canyoneering by himself. When a boulder fell from above, pinning his arm against the canyon wall, he had to make some very tough decisions about what mattered most.

Stay where he was and he would die within days of shock and thirst. His only option was to pull out a pocketknife and hack off his pinned arm one little bit at a time, spacing the timing of the cuts so he could withstand the excruciating pain. Aron was inspired to survive by a dream he had during his calamity. In essence, he saw his future and in doing so he knew he had to survive so he could live out the life he was meant to own.

This is a small smattering of the inspirational stories I read to survive my own trials. Each story was packed with daunting odds, guts, sheer determination and an intense desire to live. In each tale, the protagonist faced insurmountable odds. With no hope to be found, each of these stories could easily have ended in tragedy, and yet they did not. I ascribe this to the strength and resilience of the human spirit. Where there is a will, there is always a way through the rock and the hard place.

Plus, if you knew you were going to die...wouldn't you try? What afterall, did these people have to lose? Everything...that is the answer I come up with...they stood to lose it all.

That's why I admire the people hijacked on Flight 93, headed for the White House. Once the terrorists made themselves known, it probably didn't take long for the innocent passengers to realize they were almost certainly going to die. Rather than freezing in fear, these brave people used their last moments to make a difference for our country, forcing the plane to crash in an empty Pennsylvania field rather than hitting our nation's capital.

Attitude is everything and there's a lot to be said for sacrifice too. We are a country comprised of amazing people. Harness that energy and how could we fail to tackle the climate change problem successfully?

Take a look at the traits of these people and you have a recipe for overcoming any obstacle life can toss your way. Hope, faith, teamwork, determination and true grit, these are the things upon which America was built.

Climate change is America's rock and the hard place. It is time for us to show the world what we are truly made of. Consider it a spiritual quest. The stakes will never be higher and failure is certainly not an option.

What, afterall, do we stand to lose?

Absolutely everything.

THE BIGGEST LOSERS

Concern for our National Parks was the primary driver for writing this book, but as I delved more deeply into the subject, a far bigger and scarier apprehension arose.

Our changing climate has perched us on the brink of several natural disasters.

Not only will droughts whither some areas, causing crops to fail but wildfires will grow in intensity, chewing up more acreage by the decade. Massive extinction events, wherein we could easily lose 1 in 5 species on earth at a minimum, could occur as creatures and plants rush to find livable habitat in unknown locations only to find there is no suitable place to call home.

In time, I do believe that the natural laws set in motion by our great Creator will heal the planet, allowing a multitude of creatures to emerge following the coming disaster. Though these creatures may not resemble the species we have come to love during our own time on earth, they will probably be intricately beautiful and amazing in their own right.

Unfortunately, the biggest losers in the climate change travesty that is unfolding all around us will likely be ourselves. Species that cannot adjust or evolve will be shaken off this planet like a cloud of irritating fleas so energy systems can rebalance, restoring harmony to God's creation.

Many of us see it coming like a brakeless train careening down a mountains side. Others have not yet grasped the reality of what is unfolding and will be blindsided when the storm strikes. A third and far more sinister category of people also exists.

These people know full well that climate change is a clear and present danger but are interested in holding onto their personal agendas with steel-clenched fists while stockpiling wealth and a false sense of energy security. Meanwhile the earth is flashing warning signals portending of the coming storm. Consciously ignoring environmental signals will not lessen the approaching damage nor the chaos soon to be unleashed upon all of humanity.

History will judge this latter group harshly for there was a knowing, a hiding, and a complicitness embedded in their decision making process. How we live our lives matters immensely. What we choose to fight for defines our character. When we leave this planet, I believe we are given a chance to review the course of our lives. Perhaps no one will judge us more harshly than our very own hearts.

We are at a major crossroad and selecting a path will have dire consequences, ones that will impact our children and the very face of the planet. Bumbling along is an option, but I think we are being called to take a higher path.

We are the generation that can become the hope and light of the world, but our power is in the NOW and we must choose to act.

Wait much longer and we don't stand a chance.

SOURCES OF INFORMATION

Chapter 1: A Call to Action

Flannery, Tim (2001) The Weather Makers: How Man Is Changing the Climate and What It Means for Life on Earth

https://www.washingtonpost.com/politics/trump-scales-back-two-huge-national-monuments-in-utah-drawing-praise-and-protests/2017/12/04/758c85c6-d908-11e7-b1a8-62589434a581_story.html?noredirect=on&utm_term=.c0bc40dc747d

Chapter 2: In the Realm of Giants

Genome Project Aims to Restore Health of Redwood and Giant Sequoia Forests. (2017, October 20). Retrieved from http://blogs.ucdavis.edu/egghead/2017/10/20/

ge(nome-project-aims-restore-health-redwood-giant-sequoia-forests

Szalay, J (2017, May 04). Giant Sequoias and Redwoods: The Largest and Tallest Trees. Retrieved from https://www.livescience.com/39461-sequoias-redwood-trees.htm.

(2016, December 06). How Climate Change is Affecting Our Redwood Forests. Retrieved from https://thebolditalic.com/will-climate-change-kill-the-redwoods-la2c33d2dad7

Past, Present and Future of Redwoods: A Redwood Ecology and Climate Symposium
https://www.savetheredwoods.org/wp-content/uploads/RCCI-Symposium-2013-Abstracts.pdf

Climate Change in the Sierra Nevada, California's Water Future, 2018, Retrieved from
https://www.ioes.ucla.edu/wp-content/uploads/UCLA-CCS-Climate-Change-Sierra-Nevada.pdf

Hansman, Heather, Congress moves to give away national lands, discounting billions in revenues, Retrieved from:
https://www.theguardian.com/environment/2017/jan/19/bureau-land-management-federal-lease

Chapter 3: Our Acidic Oceans

Anderson, Robert T., and William H. Rodgers. Ocean Acidification: Understanding the Other Climate Crises: a Handbook on the Development of Ocean Acidification Science, Policy, and the Law. University of Washington School of Law, 2016.

Pearce, F. (2008) With Speed and Violence: Why scientists fear tipping points in climate change. Boston: Beacon Press.

Goodell, J. (2018). Water Will Come: Rising seas, sinking cities, and the remaking of the civilized world. S.l: Back Bay Books Little Brn.

Staletovich, J. (n.d). Beyond the high tides, South Florida water is changing. Retrieved from: http://www.miamiherald.com/news/local/environment/article41416653.html

99% of These Sea Turtles Are Turning Female-Here's Why. (2018, January 08). Retrieved from https://news.nationalgeographic.com/2018/01/australia-green-sea-turtles-turning-female-climate-change-raine-island-sex-temperature/

National Parks Traveler, (n.d.). Retrieved from: https://www.nationalparkstraveler.org/2016/06/new-research-shows-threats-coral-reefs-world-over

How the World Passed a Carbon Threshold and Why It Matters. (n.d.). Retrieved from https://e360.yale.edu/features/how-the-world-passed-a-carbon-threshold-400ppm-and-why-it-matters

Gore, A. (2017). An Inconvenient Sequel: Truth to Power. Emmaus, PA: Rodale.

Chapter 4: Statue of Liberty

The Immigrant's Statue, (n.d.). Retrieved from https://www.nps.gov/stli/learn/historyculture/the-immigrants-statue.htm

Hurricane Sandy Recovery. (n.d.). Retrieved from https://www.nps.gov/stli/after-hurricane-sandy.htm

How Global Warming Made Hurricane Sandy Worse. (2012, November 01). Retrieved from http://www.climatecentral.org/news/how-global-warming-made-hurricane-sandy-worse-15190

History & Culture. (n.d.). Retrieved from https://www.nps.gov/stli/learn/historyculture/index.htm

Hurricane Sandy Fast Facts. (2017, October 19). Retrieved from https://www.cnn.com/2013/07/13/world/americas/hurricane-sandy-fast-facts/index.html

Amadeo, K. (n.d.). How Bad Was Hurricane Sandy? Retrieved from https://www.thebalance.com/hurricane-sandy-damage-facts-3305501

Impacts of Hurricane Sandy and the Climate Change Connection, Christina DeConcini and Forbes Tompkins, Retrieved from https://www.wri.org/sites/default/files/pdf/sandy_fact_sheet.pdf

Global Warming and Hurricanes, 2018, Retrieved from https://www.gfdl.noaa.gov/global-warming-and-hurricanes/

On Estimates of Historical North Atlantic Tropical Cyclone Activity, Gabriel A. Vecchi and Thomas R. Knutson, 2007, Retrieved from https://www.gfdl.noaa.gov/bibliography/related_files/gav0802.pdf

Hurricanes and Climate Change, from Union of Concerned Scientists, https://www.ucsusa.org/global-warming/science-and-impacts/impacts/hurricanes-and-climate-change.html

Chapter 5: The Preciousness of Water

Meyer, R. (2017, January 18). Could Scott Pruitt Have Fixed Oklahoma's Earthquake Epidemic? Retrieved from https://www.theatlantic.com/science/archive/2017/scott-pruitt-and-oklahomas-manmade-earthquakes/513437

Zinke says a third of Interior's staff is disloyal to Trump and promises huge changes, Darryl Fears and Juliet Eilperin, 2017, Retrieved from
https://www.washingtonpost.com/news/energy-environment/wp/2017/09/26/zinke-says-a-third-of-interiors-staff-is-disloyal-to-trump-and-promises-huge-changes/?utm_term=.1543028db168

Without fanfare, oil companies just received a tax break on New Year's Day, Juliet Eilperin and Dino Grandoni, 2018, Retrieved from
https://www.washingtonpost.com/news/energy-environment/wp/2018/01/05/republicans-allowed-a-tax-on-oil-companies-to-expire-and-almost-nobody-noticed/

Chickasaw National Recreation Area, General Management Plan, 2008, Retrieved from
https://www.nps.gov/chic/learn/management/upload/ChickasawGMP_part1.pdf

Water resources management plan - Chickasaw National Recreation Area, Oklahoma, Wikle, T and Nicholl, M, 1998, Retrieved from
https://archive.org/details/waterresourcesma00wikl

Fracking by the Numbers. (2013, October 03). Retrieved from https://environmentamerica.org/reports/ame/fracking-numbers

Analysis of Hydraulic Fracturing Fluid Data from the FracFocus Chemical Disclosure Registry 1.0, 2015, Retrieved from https://www.epa.gov/sites/production/files/2015-03/documents/fracfocus_analysis_report_and_appendices_final_032015_508_0.pdf

What Chemicals are Used. (n.d.). Retrieved from https://fracfocus.org/chemical-use/what-chemicals-are-used

Dangerous Fracking Chemicals. (n.d.). Retrieved from http://frackinginjurylaw.com/dangerous-fracking-chemicals/

GasLand, (n.d.). Retrieved from https://dvd.netflix.com/MovieGasLand/70129353

Fishman, C. (2012) The Big Thirst: The secret life and turbulent future of water. New York: Free Press.

National Geographic Society, (2013, March 18). How Hydraulic Fracturing Works. Retrieved from https://www.nationalgeographic.org/media/how-hydraulic-fracturing-works/

Oklahoma is laboratory for research on human-induced earthquakes. (n.d.). Retrieved from https://phys.org/news/2017-04-oklahoma-laboratory-human-induced-earthquakes.html

What Should We Know about Human-Induced Earthquakes? (2017, October 19). Retrieved from https://www.seismosoc.org/news/know-human-induced-earthquakes/

Oklahoma governor asks the state to pray for oil. (n.d.). Retrieved from https://thinkprogress.org/oklahoma-asks-residents-to-pray-for-oil-6af9d74c7d40/

Brodwin, E. (2018, February 02), Parts of Oklahoma now have the same earthquake risk as California – and a new study found a scarily direct link to fracking. Retrieved from http://www.businessinsider.com/earthquakes-fracking-oklahoma-research-2018-2

Hughes, T. (2016, January 13). Oklahoma hit with 70 quakes in a week. Retrieved from https://www.usatoday.com/story/news/2016/01/07/small-earthquakes-shaking-oklahoma-blamed-deep-injection-wells/78421444/

Information by Region-Oklahoma. (n.d.). Retrieved from https://earthquake.usgs.gov/earthquakes/byregion/oklahoma.php

Whitaker, B. (2016, May 08). Oklahoma's rise in quakes linked to man-made causes. Retrieved from https://www.cbsnews.com/news/60-minutes-oklahoma-rise-in-quakes-linked-to-man-made-causes/

Proximity to Natural Gas Wells and Reported Health Status: Results of a Household Survey in Washington

County, Pennsylvania. (n.d.). Retrieved from https://ehp.hiehs.nih.gov/1307732/

Allegheny County Health Department. (n.d.). ACHD – Unconventional Well Monitoring Program – Scientific Research. Retrieved from http://www.achd.net/shale/reasearch.html

Wyoming's smog exceeds Los Angeles' due to gas drilling, Koch, Wendy, Retrieved from https://content.usatoday.com/communities/greenhouse/post/2011/03/wyomings-smog-exceeds-los-angeles-due-to-gas-drilling/1

Endocrine-Disrupting Chemicals: An Endocrine Society Scientific Statement, Evanthia Diamanti-Kandarakis, 2009, Retrieved from **https://www.ncbi.nlm.nih.gov/pmc/articles/PMC2726844/**

Potential Health and Environmental Effects of Hydrofracking in the Williston Basin, Montana. (2018, April 05). Retrieved from https://serc.carleton.edu/NAGTWorkshops/health/case_studies/hydrofracking_w.html

Fischetti, M. (2012, January 20). Fracking Would Emit Large Quantities of Greenhouse Gases. Retrieved from https://www.scientificamerican.com/article/fracking-would-emit-methane/

Schnurr, R. (2017, November 18). The Oil Pipelines Putting the Great Lakes at Risk. Retrieved from http://beltmag.com/oil-pipelines-great-lakes-risk

Lead-Laced Water In Flint: A Step-By-Step Look At The Makings Of A Crisis, Merrit Kennedy, 2016, Breaking News from NPR, Retrieved from https://www.npr.org/sections/thetwo-way/2016/04/20/465545378/lead-laced-water-in-flint-a-step-by-step-look-at-the-makings-of-a-crisis

EPA's Study of Hydraulic Fracturing for Oil and Gas and Its Potential Impact on Drinking Water Resources, (2017, January 18). Retrieved from https://19january2017snapshot.epa.gov/hfstudy_html

Goodell, J. (2017, July 27). How Scott Pruitt Is Gutting the EPA on Behalf of the Fossil-Fuel Industry. Retrieved from https://www.rollingstone.com/politics/features/scott-pruitt-is-gutting-the-epa-serving-fossil-fuel-industr-w494156

Board, T.T. (2017, September 05). How President Trump and the EPA's Scott Pruitt are making America's environment deadly again. Retrieved from http://www.latimes.com/opinion/editorials/la-ed-trump-epa-environment-pruitt-20170905-story.html

The Fragile Ground Beneath 66 Million Barrels of Oil. (2017, July 06). Retrieved from https://www.insidescience.org/news/fragile-ground-beneath-66-million-barrels-oil

Opinion, The Halliburton Loophole. (2009, November 02). Retrieved from https://www.nytimes.com/2009/11/03/opinion/03tue3.html

Chapter 6: Mystery on the Markagunt Plateau

deBuys, W. (2011), A Great Aridness: Climate Change and the Future of the American Southwest

Disturbance, structure, and composition: Spruce beetle and Engelmann spruce forests on the Markagunt Plateau, Utah by RJ DeRose, JN Long, (2007)

Climate Factors Associated with Historic Spruce Beetle (Coleoptera: Curculionidae) Outbreaks in Utah and Colorado by Elizabeth G. Hebertson and Michael J. Jenkins, in Environmental Entomology, Vol. 37, Issue 2: Pages 281-292. April, 2008.

Spruce Beetle in the Rockies, J.M. Schmid and R.H. Frye, 1977

Diapause and overwintering of two spruce bark beetle species, Martin Schebeck, et al Retrieved from https://www.ncbi.nlm.nih.gov/pmc/articles/PMC5599993/

Spruce Beetle Biology, Ecology and Management in the Rocky Mountains: An Addendum to Spruce Beetle in the Rockies, Michael J. Jenkins, et al. Retrieved from http://

citeseerx.ist.psu.edu/viewdoc/download?doi=10.1.1.885.4300&rep=rep1&type=pdf

Spruce Beetles, USDA Forest Service, Retrieved from https://www.fs.usda.gov/Internet/FSE_DOCUMENTS/stelprdb5303039.pdf

Climate change and bark beetles of the western United States and Canada: Direct and indirect effects, USDA Forest Service, Retrieved from https://www.fs.usda.gov/treesearch/pubs/36133

Confronting Climate Change in New Mexico, Retrieved from https://www.ucsusa.org/sites/default/files/attach/2016/04/Climate-Change-New-Mexico-fact-sheet.pdf

Chapter 7: The Big Thaw is On

1988 Fires, National Park Service, Retrieved from https://www.nps.gov/yell/learn/nature/1988fires.htm

The Status of Our Scientific Understanding of Lodgepole Pine and Mountain Pine Beetles – A Focus on Forest Ecology and Fire Behavior https://static.colostate.edu/client-files/csfs/pdfs/LPP_scientific-LS-www.pdf

Diversity, complexity and interactions: an overview of Rocky Mountain forest ecosystems, James N. Long, Retrieved from https://tinyurl.com/ybzee73w

Climate Change, Yellowstone National Park, Retrieved from https://www.nps.gov/yell/learn/nature/climate-change.htm and https://www.nps.gov/yell/learn/nature/global-climate-change.htm

Yale Climate Connections, Retrieved from https://www.yaleclimateconnections.org/2016/11/yellowstone-and-climate-change/

How Important is Whitebark Pine to Grizzly Bears?, Cecily M. Costello, et al, Retrieved from https://www.nps.gov/yell/learn/how-important-is-whitebark-pine-to-grizzly-bears.htm

Climate Has Led to Beetle Outbreaks in Iconic Whitebark Pine Trees, U.S. Geological Survey, Retrieved from https://nccwsc.usgs.gov/content/climate-has-led-beetle-outbreaks-iconic-whitebark-pine-trees

Soul Mates: Nutcrackers, Whitebark Pine, and a Bond that Holds an Ecosystem Together, Gustave Axelson, (2015) Retrieved from https://www.allaboutbirds.org/soul-mates-nutcrackers-whitebark-pine-and-a-bond-that-holds-an-ecosystem-together/

Better Know a Bird: The Clark's Nutcracker and Its Obsessive Seed Hoarding, Lesley Evans Ogden,

Retrieved from https://www.audubon.org/news/better-know-bird-clarks-nutcracker-and-its-obsessive-seed-hoarding

For Yellowstone and America, Climate Changes Bring Our Moment of Truth, Todd Wilkinson, Mountain Journal (2018), Retrieved from http://mountainjournal.org/will-climate-change-destroy-yellowstone

Whitlock C, Cross W, Maxwell B, Silverman N, Wade AA. 2017. 2017 Montana Climate Assessment. Bozeman and Missoula MT: Montana State University and University of Montana, Montana Institute on Ecosystems. 318 p. doi:10.15788/m2ww8w. Retrieved from http://montanaclimate.org

Ecological Implications of Climate Change on the Greater Yellowstone Ecosystem, Yellowstone Science, Retrieved from https://www.nps.gov/yell/learn/upload/Accessible-PDF-prepared-for-WEB-of-Yellowstone-Science-23-1.pdf

Chapter 8: The Moose on Isle Royale

Rapid Climate Changes Turn North Woods into a Moose Graveyard, Daniel Cusick, Scientific American, Retrieved from https://www.scientificamerican.com/article/rapid-climate-changes-turn-north-woods-into-moose-graveyard/

https://www.twincities.com/2016/01/27/new-dnr-data-gives-hints-on-whats-killing-minnesotas-moose/

The Shrinking Moose of Isle Royale, Allison Mills, 2017, Retrieved from https://www.mtu.edu/news/stories/2017/december/shrinking-moose-isle-royale.html

Extreme inbreeding likely spells doom for Isle Royale Wolves, Christine Mlot (2016), http://www.sciencemag.org/news/2016/04/extreme-inbreeding-likely-spells-doom-isle-royale-wolves

Isle Royale likely down to 1 wolf – here's why it's a big problem, Keith Matheny, (2017), Detroit Free Press, Retrieved from https://www.freep.com/story/news/local/michigan/2017/12/04/one-wolf-isle-royale-pack/902023001/

Herbivory, climate change and the future landscape of Isle Royale National Park: developing an herbivory monitoring program to adaptively manage the park's terrestrial and aquatic ecosystems, Retrieved from http://seas.umich.edu/academics/resources/capstone/herbivory_climate_change_and_future_landscape_isle_royale_national_park

Chapter 9: A Sacred Trust

The Facts on Oil and Gas Drilling in National Parks, Nicholas Lund, 2017, Retrieved from https://www.npca.

org/articles/1471-the-facts-on-oil-and-gas-drilling-in-national-parks

Trump wants to make it easier to drill in national parks. We mapped the 42 parks at risk. Sarah Frostenson, 2017, https://www.vox.com/science-and-health/2017/4/20/15272642/trump-drill-oil-gas-national-parks-map

Presidential Executive Order on Promoting Energy Independence and Economic Growth, 2017, Retrieved from https://www.whitehouse.gov/presidential-actions/presidential-executive-order-promoting-energy-independence-economic-growth/

National parks are the real losers in Trump's budget and infrastructure proposals, Jenny Rowland, 2018, Retrieved from https://thinkprogress.org/national-parks-trump-infrastructure-budget-f0530e5fa7c4/

National Parks Affected by 9B Rules, Nicholas Lund, 2017, Retrieved from https://www.npca.org/resources/3190-national-parks-affected-by-9b-rules

9B Rulemaking, Retrieved from https://www.nps.gov/subjects/energyminerals/9b-rulemaking.htm

Trump Suggests We Just Drill Everywhere, Yessenia Funes at earther.com, 2018, Retrieved from https://earther.com/trump-suggests-we-just-drill-everywhere-1821781890

Regulatory Status of Non-Federal Oil and Gas Wells in Units of the National Park Service, Retrieved from https://www.nps.gov/subjects/energyminerals/upload/nps48_recommendations_20140507_FINAL.pdf

Zinke: One-third of Interior Employees Not Loyal to Trump, Matthew Daly, 2017, Retrieved from https://apnews.com/570c910d21be41869f76d45a2c55c359

Police deploy water hoses, tear gas against Standing Rock protesters, Joshua Barajas, 2016, Retrieved from https://www.pbs.org/newshour/nation/police-deploy-water-hoses-tear-gas-against-standing-rock-protesters

Hansman, Heather, Congress moves to give away national lands, discounting billions in revenues, Retrieved from:

https://www.theguardian.com/environment/2017/jan/19/bureau-land-management-federal-lease

Chapter 10: Unstable Ground

Living in the Anthropocene, The Age of Humans, Barrow, Alaska: Ground Zero for Climate Change, Smithsonian.com, Bob Reiss, 2010.

The remote Alaskan village that needs to be relocated due to climate change, Chris Mooney, 2015, Retrieved from https://www.

washingtonpost.com/news/energy-environment/wp/2015/02/24/the-remote-alaskan-village-that-needs-to-be-relocated-due-to-climate-change/?utm_term=.39209ede7e9b

U.S. Climate Resilience Toolkit, Retrieved from https://toolkit.climate.gov/crt-search?query=kivalina

Climate Impacts in Alaska, Retrieved from https://19january2017snapshot.epa.gov/climate-impacts/climate-impacts-alaska_.html

Spruce beetle devastation returns to Southcentral Alaska – and moves north, Charles Wohlforth, 2017, Retrieved from https://www.adn.com/opinions/2017/08/30/spruce-beetle-devastation-returns-to-southcentral-alaska-and-moves-north/

Spruce Beetle Surges after a Long Slumber, USDA, Retrieved from https://s3.amazon.com/arc-wordpress-client-uploads/and/wp-content/uploads/2017/08/30042438/Spruce-Beetle-2017-Briefing-Paper-v1.pdf

[11] GAO (2009). Alaska Native Villages: Limited Progress Has Been Made on Relocating Villages Threatened by Flooding and Erosion

Permafrost Stores a Globally Significant Amount of Mercury, Paul F. Schuster et al, 2018, Retrieved from https://agupubs.onlinelibrary.wiley.com/doi/pdf/10.1002/2017GL075571

The rapidly-thawing permafrost is full of mercury, Zoe Schlanger, 2018, Retrieved from
https://qz.com/1203916/the-rapidly-thawing-permafrost-is-full-of-mercury-that-could-be-a-health-disaster/

The Arctic is full of toxic mercury, and climate change is going to release it, Chris Mooney, 2018, Retrieved from
https://www.washingtonpost.com/news/energy-environment/wp/2018/02/05/the-arctic-is-full-of-toxic-mercury-and-climate-change-is-going-to-release-it/?utm_term=.3c4d5e4b1924

Here's What Scientists Know About the Risk of a Massive Global Methane Release, Matt Smith, 2017, Retrieved from
https://www.seeker.com/earth/climate/heres-what-scientists-know-about-the-risk-of-a-massive-global-methane-release

Methane and Frozen Ground, Kevin Schaefer, Retrieved from
https://nsidc.org/cryosphere/frozenground/methane.html

History is Melting: How Climate Change is Destroying Arctic Archaeology Sites, by Eli Kintisch, 2016, Retrieved from https://pulitzercenter.org/reporting/history-melting-how-climate-change-destroying-arctic-archeological-sites

New Warning About Climate Change Linked to Peat Bogs, Vera Salnitskaya, 2015, Retrieved from http://siberiantimes.com/ecology/opinion/features/f0099-new-warning-about-climate-change-linked-to-peat-bogs/

North America has 32 Million Gallons of Dangerous Mercury Below Permafrost that is Melting from Global Warming, Kastalia Medrano, 2018, Retrieved from http://www.newsweek.com/32-million-gallons-mercury-north-america-climate-change-permafrost-802005

Summary of Expected Climate-Related Change in Southeast Alaska, Retrieve from http://ecoadapt.org/data/documents/SoutheastAlaskaClimateSynopsisTable.pdf

Chapter 11: Firestorm

https://www.epa.gov/climate-indicators/climate-change-indicators-wildfires

http://www.climatecentral.org/news/area-burned-by-us-wildfires-expected-to-double-by-2050-15355

Kodas, M. (2018). Megafire (1st edition).

McDonough, B. (2017). Granite Mountain (1st edition).

Intergovernmental Panel on Climate Change (2018), Retrieved from: https://report.ipcc.ch/sr15/pdf/sr15_spm_final.pdf

Rim Fire Recovery Newspapers: https://www.fs.usda.gov/detail/stanislaus/home/?cid=stelprd3814207

Chapter 12: Sexy Cigarettes, Nuclear Fallout and Fake News

https://qz.com/591411/an-american-voters-guide-to-the-dirty-truth-about-oil/

Sarah Alisabeth Fox (2014), Downwind: A People's History of the Nuclear West

An American Voter's Guide to the Dirty Truth about Oil, Leif Wenar, 2016,
https://qz.com/591411/an-american-voters-guide-to-the-dirty-truth-about-oil/

How Global Warming Made Hurricane Sandy Worse. (2012, November 01). Retrieved from http://www.

climatecentral.org/news/how-global-warming-made-hurricane-sandy-worse-15190

Meyer, R. (2017, January 18). Could Scott Pruitt Have Fixed Oklahoma's Earthquake Epidemic? Retrieved from https://www.theatlantic.com/science/archive/2017/scott-pruitt-and-oklahomas-manmade-earthquakes/513437

http://content.time.com/time/photogallery/0,29307,1848212_1777642,00.html

https://www.pinterest.com/pin/39336196715611469/

https://www.youtube.com/watch?v=IKqXu-5jw60

Downwinder stories:

https://www.youtube.com/watch?v=N2XXlO240ac
https://www.youtube.com/watch?v=y-GeccG7zOs
https://www.youtube.com/watch?v=yE-nji8-1Ko

Chapter 13: The Rock and the Hard Place

Not applicable, but visit bookstores for great reads mentioned within this chapter.

Chapter 14: Hope is Action, Action is Hope

350.org, https://globaldivestmentmobilisation.org

Climate Reality Project, www.climaterealityproject.org

Representative's phone number: 202-225-3121

John Cook's Skeptical Science, https://skepticalscience.com/posts.php?u=1

Chapter 15: The Biggest Losers

Flannery, Tim (2001) The Weather Makers: How Man Is Changing the Climate and What It Means for Life on Earth

Kolbert, Elizabeth (2014). The Sixth Extinction.

www.ingramcontent.com/pod-product-compliance
Lightning Source LLC
Chambersburg PA
CBHW020648220526
45464CB00001B/349